BUILDING AND DESIGNING TRANSISTOR RADIOS
A BEGINNER'S GUIDE

BUILDING AND DESIGNING TRANSISTOR RADIOS

A BEGINNER'S GUIDE

Written and Illustrated by
R. H. WARRING

LUTTERWORTH PRESS
GUILDFORD AND LONDON

First published 1977

ISBN 0 7188 2229 3

COPYRIGHT © 1977 R. H. WARRING

All Rights Reserved. No part of this publication
may be reproduced, stored in a retrieval system,
or transmitted, in any form or by any means,
electronic, mechanical, photocopying, recording
or otherwise, without the prior permission of
Lutterworth Press, Farnham Road, Guildford, Surrey.

*Printed in Great Britain by
Cox & Wyman Ltd
London, Fakenham and Reading*

CONTENTS

Chapter

	INTRODUCTION	*page* 7
1.	AERIAL CIRCUITS	13
2.	AERIALS AND AERIAL COUPLING	25
3.	SEMI-CONDUCTOR DIODES AND THEIR CHARACTERISTICS	32
4.	THE DIODE DETECTOR	39
5.	TRANSISTORS	48
6.	UNDERSTANDING TRANSISTOR CHARACTERISTICS	57
7.	AUDIO AMPLIFIERS	75
8.	REGENERATIVE RECEIVERS	87
9.	THE SUPERHET	93
10.	INTERSTAGE CONNECTIONS (COUPLING)	103
11.	FIELD EFFECT TRANSISTORS	111
12.	MISCELLANEOUS CIRCUITS	118
13.	CHECKING RADIO CIRCUITS	125
	INDEX	127

INTRODUCTION

ANY radio receiver consists of a number of separate 'stages', suitably interconnected, each stage representing the equivalent of an electronic 'building block'. The three basic 'blocks' are a *tuned circuit*, to extract radio signals from the ether; a *detector*, to turn these radio signals into *audio frequency* signals (or *af*); and an electro-mechanical device to turn these *af* signals into sound (either headphones or a loudspeaker).

Thus the simplest combination of 'blocks' is:

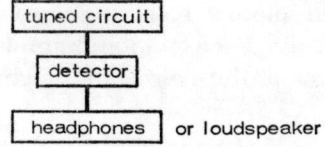

This combination has the particular limitation of providing only very low strength *af* signals—so weak, in fact, that they would certainly not work a speaker, and only give very weak signal strength in phones. Without going to more 'blocks' the only way to improve 'listening strength' is to add an external aerial.

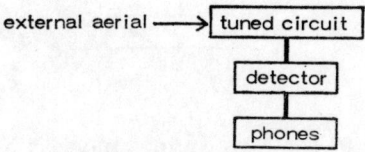

Much better results are possible by adding further 'blocks', especially one which magnifies or amplifies the signal output from the detector.

This four-block combination can work quite well, and an external aerial may not be necessary (although an advantage in some cases). Also it can be made to operate a speaker.

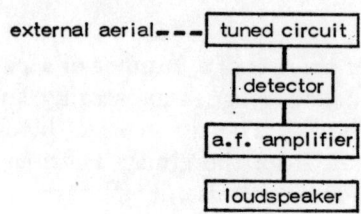

There are limits to what can be done with *af* amplifiers as these will magnify 'noise' and distortion, as well as the required *af* signals. Thus a further step, is to add another block to improve the signal quality in some way or another before detection and amplification. At this point it should also be possible to dispense with an external aerial.

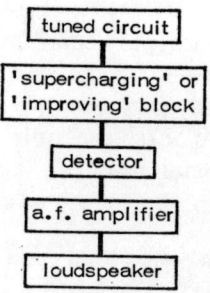

A simple solution to the 'improving' block is a regenerative circuit, which may also be designed to work as a detector as well. The more or less universal solution in modern radio receiver design, however, is the adoption of a *superhet* front end. This has the effect of extracting a

INTRODUCTION

signal frequency intermediate between the *radio frequency* (*rf*) picked up by the tuned circuit, which *intermediate frequency* (*if*) can itself be amplified before being passed to the detector stage.

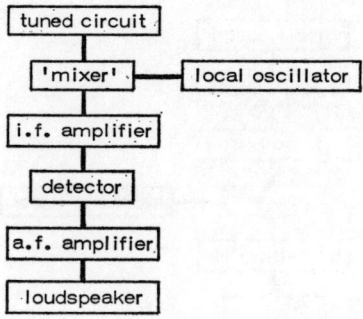

That, in fact, is as far as it is necessary to go for satisfactory reception of long wave and medium wave frequencies—or amplitude modulated (AM) broadcasts. Very High Frequency (VHF) or Frequency Modulated (FM) broadcasts, however, pose further problems for the *nature* of the *rf* signal is different, as well as the signal frequency being much higher. The latter means that the conventional form of tuned circuit for AM (usually comprising a ferrite rod aerial) is no longer suitable. Instead an external *dipole* aerial is necessary. Superhet working is more or less obligatory, so the building blocks now become:

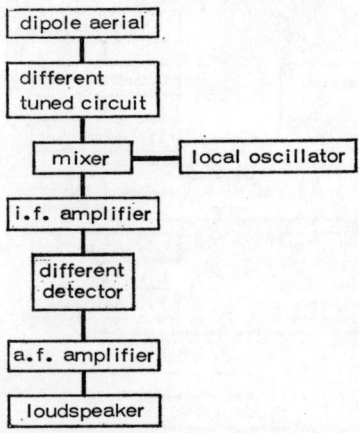

Because of the higher signal frequency at which FM is broadcast, it is also desirable to *preamplify* the incoming *rf* signal, so one more block can be added with advantage:

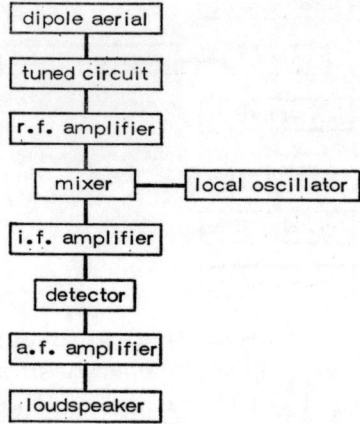

Finally, of course, you may want to build an AM/FM radio, when some of the blocks have to be duplicated, the appropriate 'blocks' being selected by a wavechange switch:

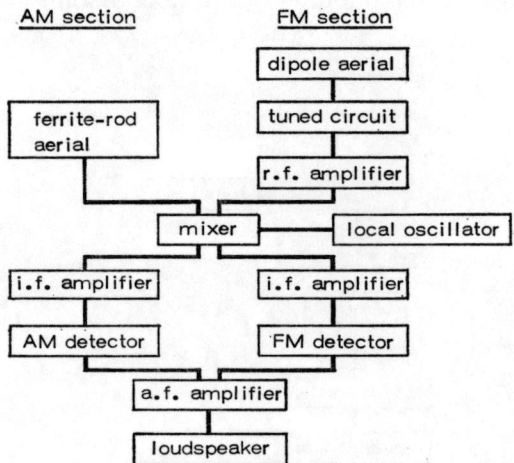

INTRODUCTION

It is the purpose of this book to describe and illustrate the design and working characteristics of the various 'blocks' from which radio receivers can be constructed. In some cases there are more or less standard designs which have evolved, e.g. for tuned circuits, AM and FM detectors, superhet mixer-oscillators, *if* amplifiers and *af* amplifiers. It would be difficult to improve on them with simple constructions. The main field for experiment is possible combinations of these various 'blocks'—and in particular the manner in which they should be connected together or *coupled* for most efficient working.

By treating all the major subjects separately it is hoped that the various design principles will have been made clear—and also how 'standard' types of 'boxes' can be connected together to make a working unit. It is also shown how additional simple circuits can be incorporated to improve the performance of any basic design.

CHAPTER 1

AERIAL CIRCUITS

THE conventional aerial circuit or *tuned circuit* comprises a variable capacitor in parallel with an inductance (physically, a coil). The frequency (f) at which such a circuit is *resonant* is given by:

$$f = \frac{1}{2\pi\sqrt{LC}}$$

f = frequency in Hz
L = inductance in henrys
C = capacitance in farads

A practical version of this formula is:

$$f = \frac{160000}{\sqrt{LC}}$$

f = frequency in kHz
where L = inductance in microhenrys (μH)
C = capacitance in picofarads (pF)

It can be noted here that a practical inductance will also have a certain amount of *resistance*, and so the equivalent circuit is as shown in the second diagram in Fig. 1. The presence of such resistance does not affect the resonant *frequency* of the circuit but only the *sharpness* of the resonance of the circuit. This controls the *quality factor* of the tuned circuit (*see later*). The practical capacitance also has a certain amount of resistance, but this is normally negligible except at very high frequencies (30 MHz and above).

By fixing one component value (e.g. inductance) and making the other variable (e.g. capacitance) it is possible to adjust or tune the circuit over a range of resonant frequencies. Theoretically, on this basis, it is possible to design a tuned circuit to cover the whole range of broadcast frequencies from the 'top' (wavelength) end of the long wave band (30 kHz) to the 'bottom' (wavelength) end of the VHF band

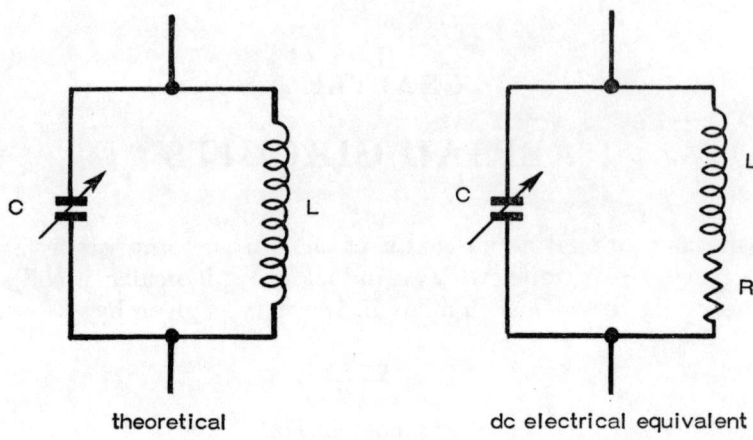

Fig. 1 Capacitance (C) and inductance (L) forming a tuned circuit.

(300 MHz). This is not a realistic solution, and so tuned circuits are designed separately to cover individual broadcast bands, e.g.

Long wave (or low frequency)—30–300 kHz
Medium wave (or medium frequency)—300–3000 kHz
Short wave (or high frequency)—3–30 MHz
VHF—30–300 MHz

In practice, tuned circuits are designed to cover the actual spread of broadcast stations operating in these bands, e.g.

Long wave—50–150 kHz
Medium wave—500–1500 kHz
Short wave—1·8–28 MHz
VHF—88–100 MHz

A significant fact is that the actual frequency range covered increases considerably with decreasing wavelength of these broadcast bands, e.g.

Long wave—range covered 100,000 Hz.
Medium wave—range covered 1,000,000 Hz.
Short wave—range covered 26,000,000 Hz.
VHF—range covered 12,000,000 Hz.

AERIAL CIRCUITS

This makes the design of aerial circuits increasingly critical from long wave upwards (in frequency). Again, in practice, this means that home-made coils are seldom suitable for other than simple long wave and medium wave receivers. Even then, proprietary coils almost invariably give better results because of the better *quality factors* (or Q) achieved. Nevertheless it is interesting to cover the design of simple aerial coils.

The simplest type of inductance is an open coil of insulated wire wound on a former of insulating material, or it can even be self-supporting if the wire is thick enough. In the latter case the coil is wound on a mandrel and then slid off, being mounted on the wire ends (Fig. 2).

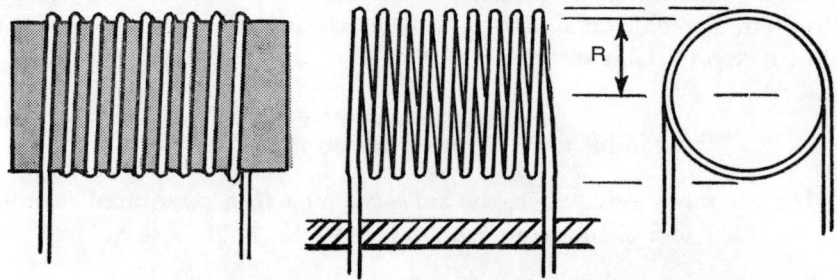

Fig. 2 Air-cored tuning coils. Coil on left is wound on a former.

The *inductance* of such a coil is found as follows:

$$\text{inductance, microhenrys} = \frac{R^2 \times N^2}{9R + 10L}$$

where R is the mean radius of the coil in inches
L is the length of the coil in inches
N is the number of turns

The effect of wire *diameter* is not significant, provided the coil diameter is reasonably large, i.e. 1 in. (25 mm) or more. It is therefore logical to use quite thick wire (18 s.w.g. or 16 s.w.g.) in order to minimize coil resistance.

Suppose such a coil is to be designed as the inductive component in a *medium wave* tuned circuit. The resonant frequency range required is 500 to 1500 kHz. Considering the requirements, first in terms of the product of L and C, from the resonant frequency formula:

$$LC = \frac{(160\ 000)^2}{f^2}$$

Thus at $f = 500$ kH$_2$, $$LC = \frac{(160\ 000)^2}{(500)}$$
$$= 102\ 400,\ \text{say}\ 100\ 000$$

at $f = 1400$ kHz, $$LC = \frac{(160\ 000)^2}{(1500)^2}$$
$$= 11{,}300,\ \text{say}\ 11{,}000$$

For a fixed value of inductance, maximum capacity will be required to tune to the lowest frequency, i.e. the highest calculated value of LC required. Typically available variable capacities offer a range of 0–200 pf or 0–500 pf. Choosing the 0–500 pf size, at maximum capacity:

$$L \times 500 = 100{,}000$$
or inductance required $= 200$ microhenrys.

Using the same inductance, the *minimum* capacitance required to tune to the other end of the band (1500 kHz) would be:

$$200 \times C\ min = 11{,}000$$
or $C\ min = 55$ microhenrys

Thus a 200 μH inductance would be a suitable match to a 0–500 pf capacitor to cover the range required.

To simplify the coil design we can 'guesstimate' a length of 1 in. and a coil diameter of 1 in. Inserting these values in the appropriate formula:

$$\text{inductance},\ \mu H = \frac{(0 \cdot 5)^2 \times N^2}{9 \times 0 \cdot 5 + 10}$$
$$\mu H = 0 \cdot 0172 N^2$$

Inserting the value of inductance required (200 μH) and solving for number of turns:

$$N = \frac{200}{0 \cdot 0172}$$
$$= 108\ \text{turns, or say 100 turns.}$$

AERIAL CIRCUITS

Close winding 100 turns of wire the length of 1 in. would permit the use of a maximum wire size of 0·01 in., say 36 s.w.g. To use a larger wire size it would be necessary to increase the length of the coil and re-calculate the number of turns required accordingly.

A *long-wave* coil would require *more* turns; and a short-wave coil less turns (perhaps only one or two turns).

Q Factor

The effect of *resistance* in the tuned circuit is shown in simple diagrammatic form in Fig. 3, representative of a resonant circuit. The current

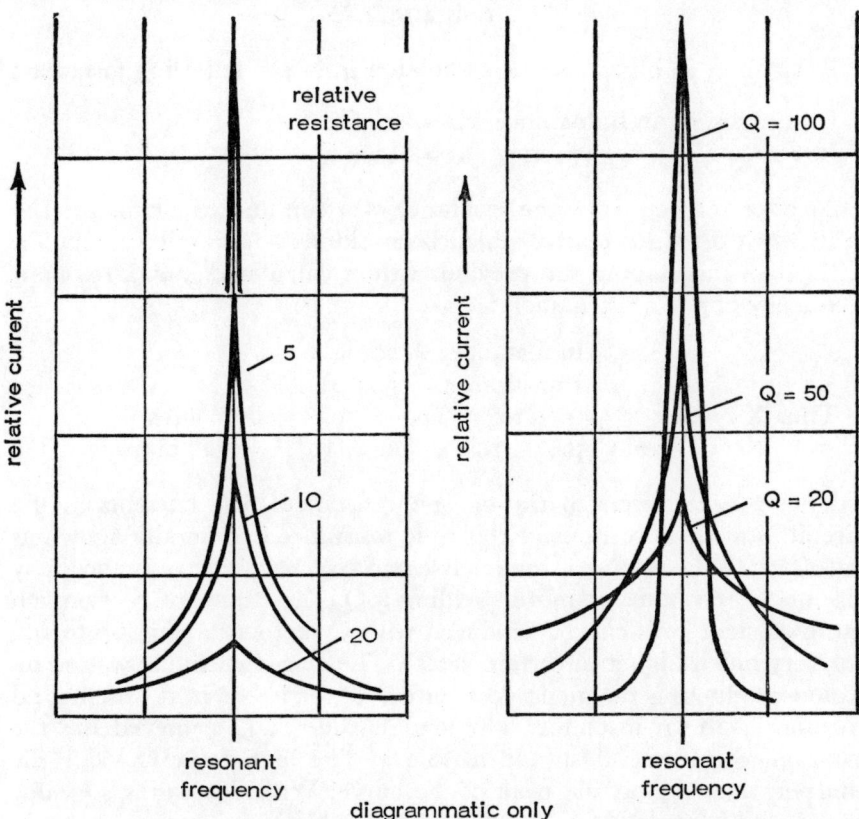

Fig. 3   The effect of resistance and Q factor on sharpness of tuning.

flowing in a resonant circuit peaks at the resonant frequency and falls off sharply on either side. The lower the *resistance* present, the *higher* the peak (more current flowing) and the sharper it is (the sharper the tuning). Resistance values shown are nominal only to illustrate this effect.

This can be put another way. The shape of the resonant curve is dependent on the respective values of the *reactance* on either the coil or capacitor and the resistance present. The ratio of the two is known as the Q factor, when

$$Q = \frac{\text{reactance (X)}}{\text{resistance (R)}}$$

Reactance, in ohms, can be calculated from the following formulas:

In the case of an inductance, $X_L = 2\pi fL$
In the case of a capacitor, $X_C = 1/2\pi fL$

At resonant frequency the reactance of a coil and capacitor are the same, so it does not matter which is considered. A simple calculation will prove this, taking the previous values calculated and a resonant frequency of 500 kHz, namely:

$$\text{inductance} = 200 \ \mu H$$
$$\text{capacitance} = 500 \ pF$$

Thus $X_L = 2\pi \times 500 \times 10^3 \times 200 \times 10^{-6} = 628$ ohms
$X_C = 1/2\pi \times 500 \times 10^3 \times 500 \times 10^{-12} = 628$ ohms

The *resistance* refers to the *dynamic* resistance to *rf* currents in the circuit, not the *dc* resistance. Dynamic resistance is generally known as *impedance*. In the case of a simple air-cored coil, dynamic resistance may rise up to 100 ohms or more, yielding a Q of less than 10. Very much more efficient coils can be produced with Q factors ranging up to 100 (or very much higher in certain cases). These are invariably wound on a non-conducting magnetic core, either of ferrite or iron dust bound together with an insulator. The actual value of Q achieved has the same effect as that illustrated in Fig. 3. The higher the Q value the sharper, and higher, the peak of the curve. With decreasing Q value the tuning becomes *broader* and the peak value is reduced as shown in Fig. 3. Sharpness of tuning is always desirable in radio receivers as it

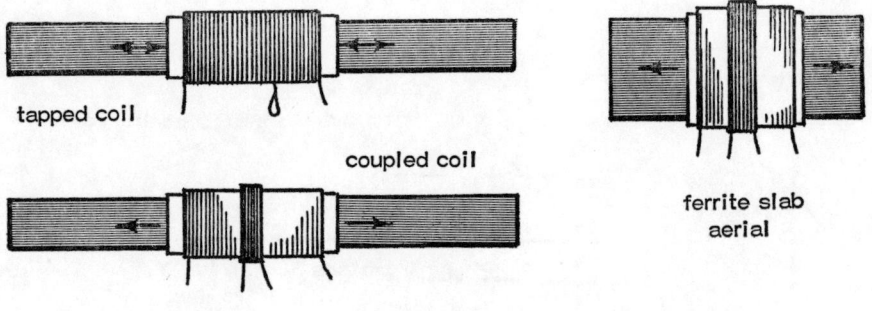

Fig. 4 Basic forms of aerial coils wound on a ferrite rod or ferrite slab.

gives good *selectivity*, or the ability to separate one station from another when the two are closely spaced on the frequency band, but see later under *Modulated Signals*.

Ferrite Rod Aerials

In simple terms introduction of a ferrite or similar magnetic material core to a coil greatly increases its inductance. This means that the coil can be made much more compact thus requiring less wire length and less resistance. A smaller wire size can also be used without introducing excessive resistance. Unfortunately no simple formulas apply for the design of such coils, for the size and number of turns required are related to the size and type of magnetic core material used. They are therefore designed on empirical or semi-empirical lines, the latter using charts related to the specific material properties.

Simple coils of this type are wound on standard sizes of ferrite rod or ferrite slab, either as tapped coils or inductively coupled coils—Fig. 4. Some design data are given in Table 1 (end of chapter).

One other advantage offered by 'cored' coils is that their inductance can be varied, if necessary, by altering the position of the coil on the core. This can be a very useful feature for adjusting the resonant frequency range of a tuned circuit independent of the variable capacitor, e.g. for setting up or 'trimming' purposes. Once adjusted in this way, the coil is then usually locked to the core (e.g. with adhesive or hard wax) to ensure that it remains at a fixed inductance.

The 'Q' of a coil can be further influenced by special forms of

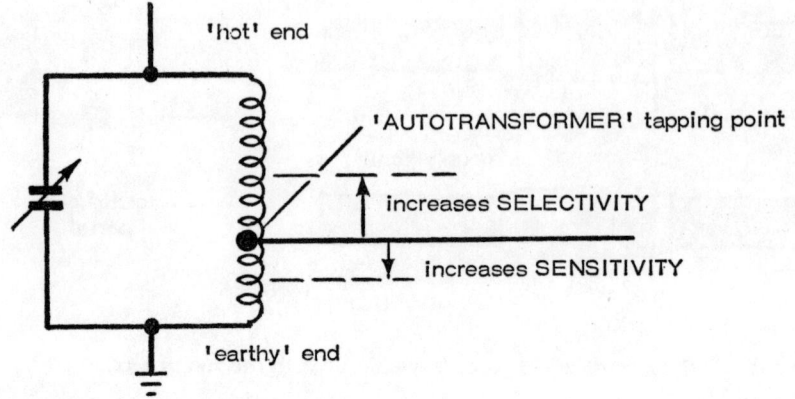

Fig. 5 The tapping point on the aerial coil affects selectivity and sensitivity.

windings, which can only be tackled successfully by coil winding machines. This is another reason why proprietary aerial coils are (almost) invariably better than home-made coils.

Selectivity and Sensitivity

Selectivity or the ability to tune sharply has already been explained. *Sensitivity* is the ability to amplify the very weak *rf* signals received in the aerial circuit to a practical output level, both as regards signal strength and depth of modulation. It is thus just as important as selectivity for satisfactory receiver performance. Both selectivity and sensitivity increase with increasing Q, but the two are not necessarily compatible.

As a basic example take the simple form of tuned circuit with a tapping point on the coil for connecting directly to the detector stage, as in Fig. 5. The end of the coil connected to the earth of the circuit is referred to as the 'earthy' end. The tapping point on a coil normally comes about one third the length (number of turns) from this end of the coil. If this tapping point is moved up towards the other or 'hot' end of the coil, this will have the effect of increasing *selectivity* but reducing signal strength or *sensitivity*. Conversely, moving the tapping point towards the 'earthy' end of the coil will reduce selectivity, but increase sensitivity. This, in fact, is one way of adjusting the selectivity and sensitivity of a simple tuned circuit of this type, the aim being to

AERIAL CIRCUITS

arrive at an optimum tapping point which gives the best possible compromise between selectivity and sensitivity. In practice this does not necessarily mean physically altering the tapping point. With a simple coil it is more practical to add turns at one end and remove turns from the other to 'shift' the tapping point.

Modulated Signals

It is possible to increase the Q attainable from coils to very high levels, using some form of positive feedback to neutralize, or partially neutralize, resistance losses. This would seem an ideal arrangement to get very sharp tuning. However, tuning can be made too sharp for a radio receiver as, it has to accept not a single frequency represented by the *rf* carrier wave, but a whole band of frequencies representing the modulated signal, otherwise some of the *af* content may be 'tuned out' or cut off. This is particularly true in the case of FM receivers, where the signal represents a *bandwidth* rather than a specific *rf* frequency. It is in this respect, both for AM and FM radio working, that the superhet receiver scores since it changes the *rf* carrier and its sidebands to a single fixed frequency, known as the *intermediate frequency*, and so selectivity can be sharply peaked.

Hearing Two Stations at once

Even the superhet receiver is not immune from interference, however, and simpler receivers considerably less so, due to what is known as *cross-modulation*. This is caused by the presence of a strong signal near to, but not at, the frequency to which the receiver is tuned. If very much stronger than the actual 'tuned' signal, it can effectively modulate that signal. In other words, although the set is tuned to a particular carrier wave frequency which has its own modulation, this carrier is now subject to modulation from the spurious signal. Thus the carrier frequency to which the set is tuned is actually carrying *two af* programmes. Hence two stations are heard simultaneously.

This is something which can readily happen on the medium wave band, especially when the set is tuned to a relatively weak station and has a much stronger station signal present close to that particular frequency. The cause may not lie in improving the selectivity of a tuned circuit so much as improving the *linearity* of a detector, for it is at this stage that the trouble will show up. Sharper tuning (i.e. better

BUILDING AND DESIGNING TRANSISTOR RADIOS

selectivity) should help, however, for the sharper the tuning the more a tuned circuit will *automatically* tend to reject any cross-modulation.

Cross modulation does not occur as a problem in the design of VHF receivers, which is why the FM band gives better reception.

Loft Aerials

The loft is the logical place to put an external aerial to improve reception in areas where radio reception tends to be poor. It can be a single dipole or a folded dipole, preferably arranged vertically so as to be non-directional, or horizontally lined up with a particular transmitting station. Its performance can be improved further with the addition of a *reflector* and *directors*. These are short lengths of wire or tube which act as 'false' aerials (or *parasitic* aerials, as they are called), with beneficial effects if they are placed in specific positions—*see* Fig. 6.

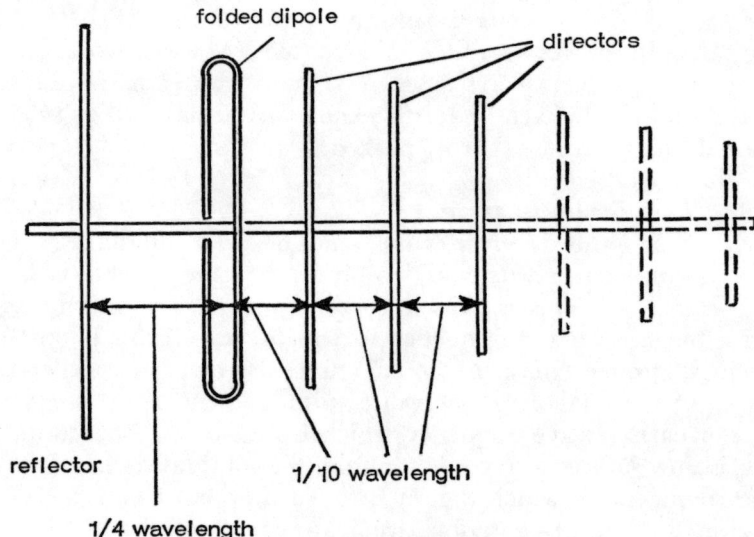

Fig. 6 The simple folded dipole aerial associated with a reflector and directors, or 'parasitic' aerials.

A reflector needs to be about the same length as a dipole or slightly longer, located a quarter of a wavelength away (e.g. about 15 in. in the case of a VHF radio aerial). This will have the effect of strengthening the signals coming from the other side and, more important still, reducing interference from the side on which the reflector is.

A false aerial placed on the opposite side to the reflector will have the effect of improving the tuning characteristics of the aerial, again provided it is the right length and in the right position. It needs to be shorter than a dipole and about one-tenth of a wavelength away from it. Its actual length and position will alter the phase relationship between the voltage and current induced in the dipole aerial. In its optimum position it will improve the tuning characteristics for a signal coming from the director-dipole direction. Further improvement may be obtained by adding still more directors, each one shorter than the previous one. This also reduces the impedance of a dipole aerial, which is an advantage for cable matching when using a folded dipole. Folded dipoles typically have an impedance of 300 ohms, although this is reduced by a reflector and director. They may still need a balancing transformer or *balun* to match a receiver aerial circuit fed by a 75-ohm coaxial cable.

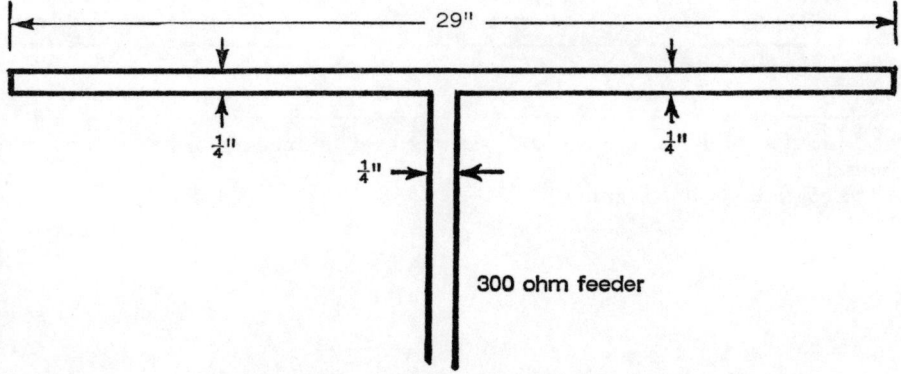

Fig. 7 Ribbon-type folded dipole wire aerial with an effective feeder impedance of 300 ohms.

A simple folded dipole can be made from a single length of insulated wire arranged in the form of an outline 'T', as shown in Fig. 7. Spacing between the wires should be ¼ in. The parallel 'leg' of the 'T' forms the feeder, with an effective impedance of 300 ohms. A simple—and inexpensive—indoor aerial of this type should be readily available from radio shops as a *ribbon* aerial (a plastic ribbon with two parallel wires incorporated in each edge of the 'ribbon' which forms the bar of the 'T', and also in the separate ribbon forming the 'leg').

Quite elaborate aerial forms consisting of a dipole and reflector and directors are used for television reception, where even higher broadcast frequencies are involved. What may appear puzzling at first is that all the units seem to be electrically connected, i.e. folded dipole, reflector and directors are all mounted on a common metal rod. In fact no electrical connection is involved as the centre of a folded dipole aerial is at zero potential.

Table 1. Examples of Medium-wave Aerial Coil Windings Matching Tuning Capacitor 0–500 pF*

Ferrite rod dia.† in.	mm	wire size s.w.g.	no. of turns	tap at	alternative coupling coil
$\frac{1}{4}''$	6.5	36 or 38	80	60 turns	16 turns
$\frac{5}{16}''$	8	36 or 38	70	55 turns	15 turns
$\frac{3}{8}''$	10	28 or 32	60	50 turns	10–15 turns
Ferrite slab† $\frac{3}{4}''$ (19 mm) × $\frac{1}{8}''$ (3 mm)		28 or 32	60	10 turns	10 turns

* Increase turns by approximately 25 per cent to match a 0–350 pF tuning capacitor.

† Length is not important.

CHAPTER 2

AERIALS AND AERIAL COUPLING

SATISFACTORY performance can usually be obtained for long wave and medium wave reception with a simple ferrite rod aerial fitted internally, although this will be very directional. The rod is usually mounted horizontally (or intended to be used in that attitude). It will produce maximum signal strength when at *right angles* to the direction of the transmitting station's aerial, with falling strength at any lesser angle, reaching a minimum when the rod is pointing *towards* the transmitting aerial. This, in fact, applies to any aerial formed from a *coil* or *loop* of wire. Maximum signal strength is received when a *loop* is *in line* with the transmitting aerial; and minimum signal (theoretically zero signal) when the loop is at right angles.

To render such an aerial non-directional it is necessary to connect it with an external aerial, which in its simplest form is a length of wire run vertically. Theoretically, at least, such an aerial would work on its own as a tuned circuit, if it was made a resonant length of $1, \frac{1}{2}, \frac{1}{4}, \frac{1}{8}$, etc. of a wavelength (*see also* Table 2 at end of chapter). This would necessitate a very high aerial to give any reasonable degree of signal amplification, and also one whose length would have to be adjusted to tune to different broadcast frequencies.

In practice, therefore, any shorter length can be used, connected to a conventional tuned circuit as in Fig. 8. The latter is then used to adjust the resonant frequency of the whole. Since the actual length of aerial wire to be used is small, compared with a wavelength, the tuned circuit can be designed to cover the normal frequency range required as in Chapter 1. The external aerial is then regarded simply as an appendage. Its addition will modify the performance of a tuned circuit to some extent. The longer—and thus the more efficient it is—the greater the modifying effect.

This is one reason why the average radio receiver fitted with an internal ferrite rod aerial normally has no provision to plug in an external aerial. Connecting an external aerial would simply mis-tune the set, which is calibrated against the adjustable tuning range provided by its own internal tuned circuit.

In other words, if an external aerial is to be used the tuning should be calibrated with external aerial attached. If it is to be provided as an additional facility, e.g. to improve reception on weak signals, then the method of coupling to the existing tuned circuit needs particular attention if mistuning is to be avoided. In the first case it is possible to connect the external aerial directly to the tuned circuit as in Fig. 8.

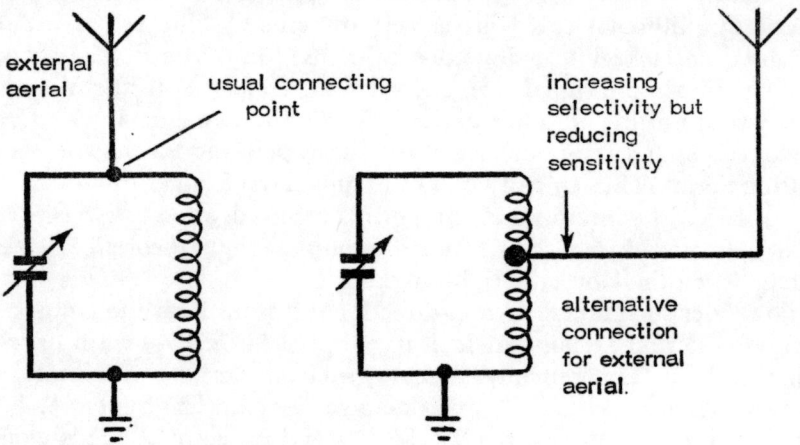

Fig. 8   External aerial connected to the tuned aerial circuit.

Normally it would be connected directly to the 'hot' end of the coil, as this should give maximum *sensitivity*. However, better *selectivity* may be produced if the connection is made lower down the coil, at the expense of sensitivity. In most practical receiver circuits, other than elementary designs, the aerial circuit is inductively coupled to the next stage of the receiver. This provides variable control of the coupling, and the possibility of better matching of the aerial to the input requirements of the following stage.

The principal trouble associated with the use of a simple wire aerial is that it can have a considerable capacitance effect. The fact the length of wire also has resistance, and thus effectively inductance, is why it can work on its own as a tuned circuit. The actual modifying effect of a wire aerial is not readily determinable, and varies both with its length and frequency of signal. It could be high enough seriously to interfere with the working of a tuned circuit.

The answer to this is surprisingly simple—just connect the external aerial via a small value series capacitor, as in Fig. 9. The effective capacitance added to the tuned circuit cannot be greater than the value of this series capacitor and so choosing a small value of 100 pf or less will ensure that the tuned circuit is not mis-tuned to any great extent. Also this method of coupling prevents the tuned circuit from being

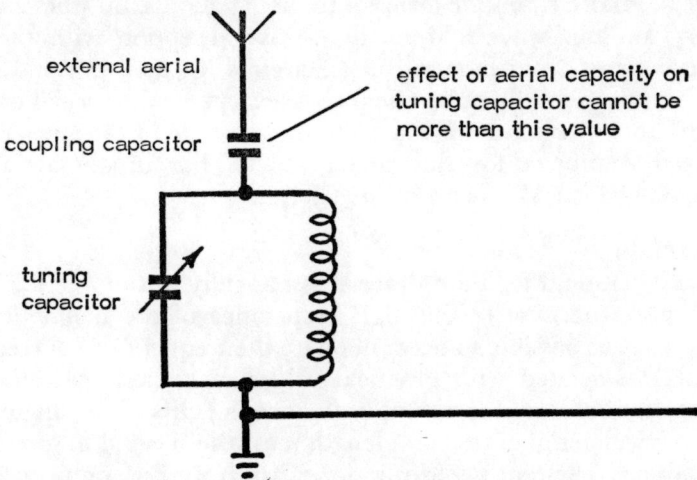

Fig. 9 Coupling capacitor between an external aerial and tuned circuits limits 'capacitance effect' of the external aerial, which tends to modify tuning.

swamped by the external aerial, i.e. the two are *under*-coupled rather than *over*-coupled. This should improve both selectivity and sensitivity and very large aerial lengths can be used without upsetting the working of a tuned circuit. On the other hand, if the tuning of the receiver is to be calibrated *with* an external aerial attached, it would be advantageous to use a higher value of coupling capacitor in order to make the external aerial more effective.

As previously noted, the greater the height of a vertical aerial the better. However the effective length can be increased by connecting a vertical wire to a horizontal wire producing an aerial in the form of a large 'T' or an 'L'. The longer the horizontal run the greater the signal strength generated in the vertical wire. The longer the length of horizontal wire the better, with the advantage that this is easier to arrange than vertical wiring. This again adds a directional quality to the area by virtue of a horizontal length which can be eliminated by running the horizontal wire through a 90-degree bend. This arrangement is quite suitable for accommodating in a loft or even in a single room. Note: the 'directional' characteristics of a horizontal *wire* aerial are the opposite to that of a coil or loop aerial. It is most effective when the wire is *in line* with the transmitting station.

There are more elaborate forms of receiving aerials, but these are not necessary for long-wave and medium-wave reception with the high efficiency offered from modern tuned circuits. In fact, internal ferrite rod aerials are generally quite adequate—except for the special requirements of short wave and VHF reception. Coverage of the former is too specialized a subject for this present book, but dipole aerials are commonplace for FM reception (*see* Chapter 1).

VHF Aerials

The conventional ferrite rod aerial is basically unsuitable for VHF/FM reception because of the high inductance of the magnetic core. This makes it impossible to accommodate the frequencies covered (88–100 MHz) associated with practical values of variable capacitors. A favourite form of aerial is a simple (telescopic) dipole. A single (telescopic) vertical aerial of resonant length may also be used in some cases.

A good aerial system is essential for satisfactory FM radio receivers. Requirements are also complicated by the fact that VHF/FM signals are both highly directional and also *polarized* (i.e. radiated in a horizontal or vertical plane). The first situation is met by aligning the dipole aerial with the broadcast station to get maximum signal strength. The second is met by aligning the aerial in the plane of polarization. Most VHF/FM signals are horizontally polarized, calling for the aerial to be aligned in a horizontal plane.

The theoretical resonant length of a dipole or *half-wave* aerial is calculated from the formula:

$$\text{Length} = \frac{492}{f} \text{ feet}$$

$$= \frac{5905}{12f} \text{ inches}$$

where f is the frequency in MHz.

AERIALS AND AERIAL COUPLING

Thus to match a nominal middle frequency in the VHF range the value of f can be taken as 95 MHz, when

$$\text{Length} = \frac{5905}{95}$$
$$= 62 \text{ inches, or say 60 inches.}$$

Strictly speaking theoretical length should be corrected for 'end effect' losses, and also the size of the wire or tube used for conductors. However since the match is only nominal to the middle of the range to be covered, and the resonant frequency of the tuned circuit will be variable, the actual length used is not that critical. Shorter lengths can also be used in multiples, e.g. 30 in. or 15 in. The vertical VHF aerial can be of any of these lengths. A horizontal dipole should have a full 60 in. length. (Fig. 10.) There are other more elaborate forms of folded dipole aerial which can be used, as mentioned in Chapter 1.

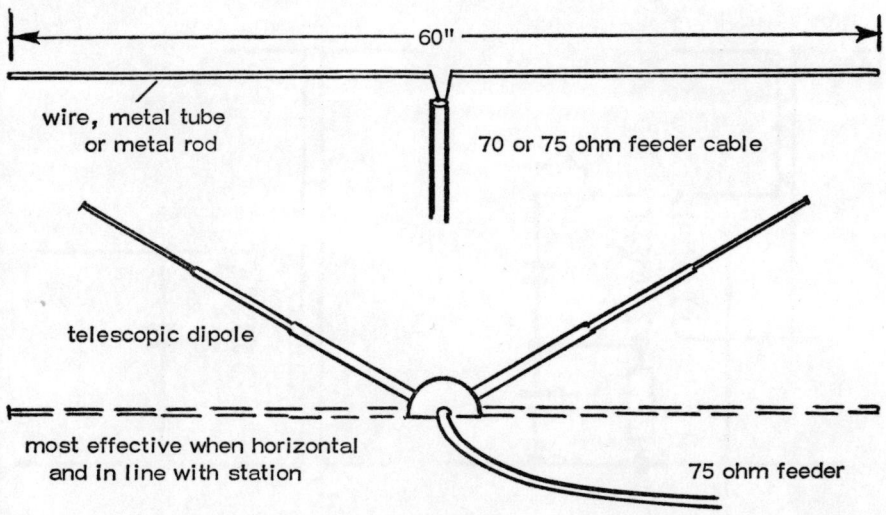

Fig. 10 The dipole aerial—the standard form of aerial used with VHF (FM) receivers.

Telescopic aerials have the advantage that a resonant length can be adjusted for optimum effect, if necessary. Also the vertical attitude of a telescopic dipole can be altered by adjusting the spread. This type of aerial can be designed either to fit on the back of a receiver or on a separate stand. It is also usually mounted so it can be adjusted directionally.

It is essential that dipole aerials be connected to the receiver tuned circuit with a properly matched cable. Dipole aerials can be taken as offering a purely resistive load for matching purposes, with low or very low reactance. Dynamic resistance or impedance is specified for various proprietary types and these are usually supplied with matching cables. A typical dipole has an impedance of about 70 ohms at the centre, and should be connected via 70–75 ohm coaxial cable, e.g. the standard type of cable used for connecting a TV aerial to its set. Folded dipoles have a higher impedance, (theoretically four times that of a simple dipole, or typically about 300 ohms), but this is reduced by the effect of a reflector and directors in the case of an aerial array of the type already described in Chapter 1.

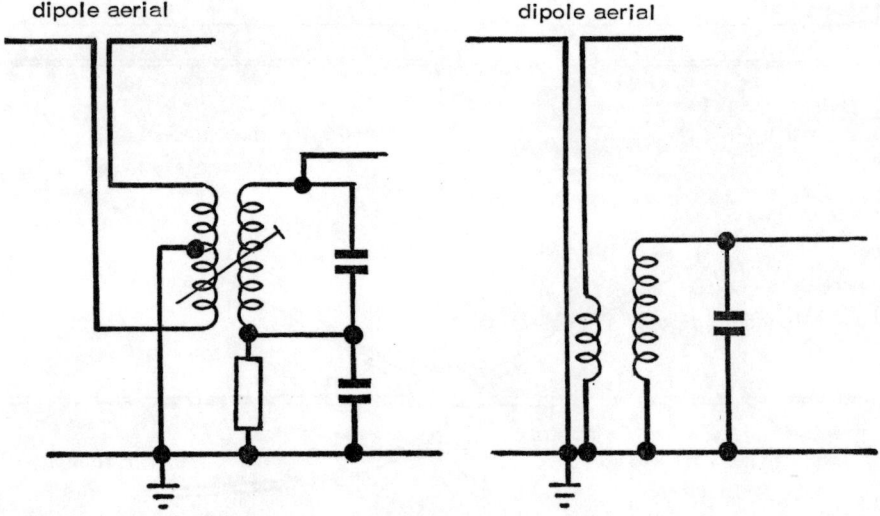

Fig. 11 Two methods of coupling a dipole aerial via impedance-matching transformers.

Since a dipole aerial involves two leads (in the feeder), it cannot be connected simply to a tuned circuit like a simple wire aerial. Instead, it is normally inductively coupled to the receiver circuit. Two such

AERIALS AND AERIAL COUPLING

methods are shown in Fig. 11. In the first illustration the dipole aerial is inductively coupled to the tuned circuit via a centre-tapped transformer, the transformer also providing the required impedance match. The second shows simpler transformer coupling (again with impedance matching) to a *preselector* circuit, which is commonly used with an FM superhet receiver.

Table 2. Resonant Aerial lengths

Radio frequency MHz	Theoretical resonant length* (metres)	½-wave inches	¼-wave inches	⅛-wave inches
80	3·75	74	37	18½
82	3·66	72	36	18
84	3·57	70	35	17½
85	3·53	69	34½	17¼
86	3·49	68	34	17
88	3·41	67	33½	16¾
90	3·33	66	33	16½
92	3·26	64	32	16
94	3·19	62	31	15½
96	3·125	61	30½	15¼
98	3·06	60	30	15

* The practical equivalent aerial length is slightly less than the theoretical value.

CHAPTER 3

SEMI-CONDUCTOR DIODES AND THEIR CHARACTERISTICS

A diode is an electronic device with *two* electrodes. Its basic action is that of a *rectifier*, i.e. it allows current to flow in one direction but not in the other. A semi-conductor diode also has the characteristics of a *capacitor*, as will be explained later.

Solid state diodes consist of a P-type semi-conductor material joined to an N-type semi-conductor material. As far as conduction of electricity is concerned, P-type material conducts by *positive* carriers (known as holes), and an N-type material by *negative* carriers, or electrons.

A simple combination of a P-type material and an N-type material joined or connected together will not work. It has to be a single crystalline element, one side of which is given P-type characteristics by 'doping' with a suitable impurity, and the other side given N-type characteristics by doping with another impurity. The two zones are bounded by a P-N junction. Mobile charges in the two materials will then tend to diffuse across this junction, cancelling each other out and forming what is in effect a neutral zone at the boundary, known as a *depletion* layer—see Fig. 12.

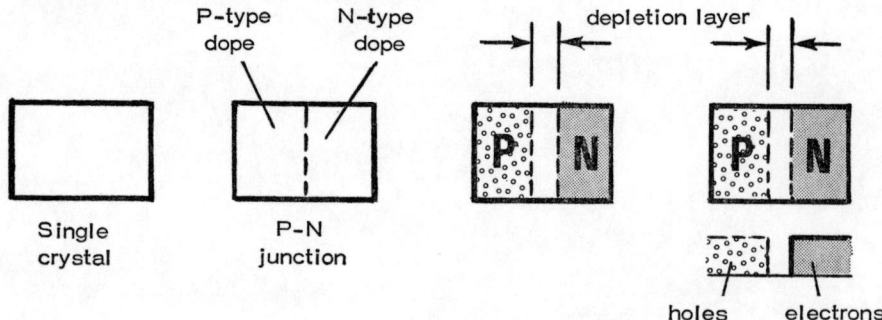

Fig. 12 Diagrammatic representation of the production of a semi-conductor diode.

This depletion layer, which is extremely thin, effectively isolates the *fixed charges* present in the crystal—positive on the N side and negative on the P side—and also prevents migration of further *mobile* charges across the junction, unless the *potential barrier* at the junction is upset by applying an external electromotive force (*emf*) or voltage to the device.

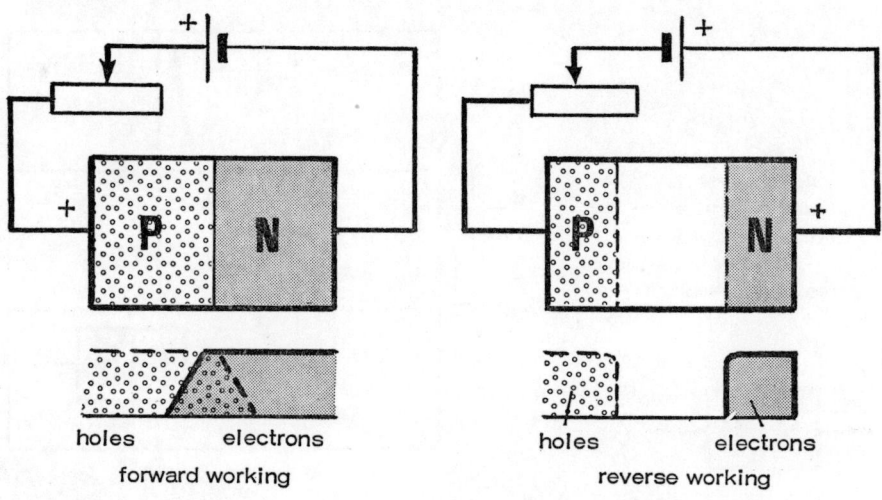

Fig. 13   The two modes of working a diode.

A 'demonstration' circuit is shown in Fig. 13, using a battery for the source of *emf* and a means of varying the voltage. In the first diagram the positive side of the battery is connected to the P side of the diode. Starting from zero and gradually increasing the voltage the first effect is to push holes from the P material towards the junction and attract electrons from the N materials towards the junction until the potential barrier is neutralized. If the *emf* is then increased further the diode will become fully conducting, with current flowing through it. In other words, applying *positive* bias to the P side of the diode has turned it into a conductor.

If *reverse bias* is applied (negative to the P side) as in the second diagram, both holes and electrons are repelled from the junction, increasing the extent of the depletion layer. In effect the junction sets up a back voltage equal to the applied voltage and a diode behaves as a non-conductor. In practice there will be a small leakage current, although this can be negligible for a majority of applications.

The typical working characteristics of a semi-conductor diode are shown in Fig. 14. Note that this presentation, which is conventional, is the opposite way round to the two diagrams in Fig. 13. In the forward direction, increasing voltage at first has no effect until the potential

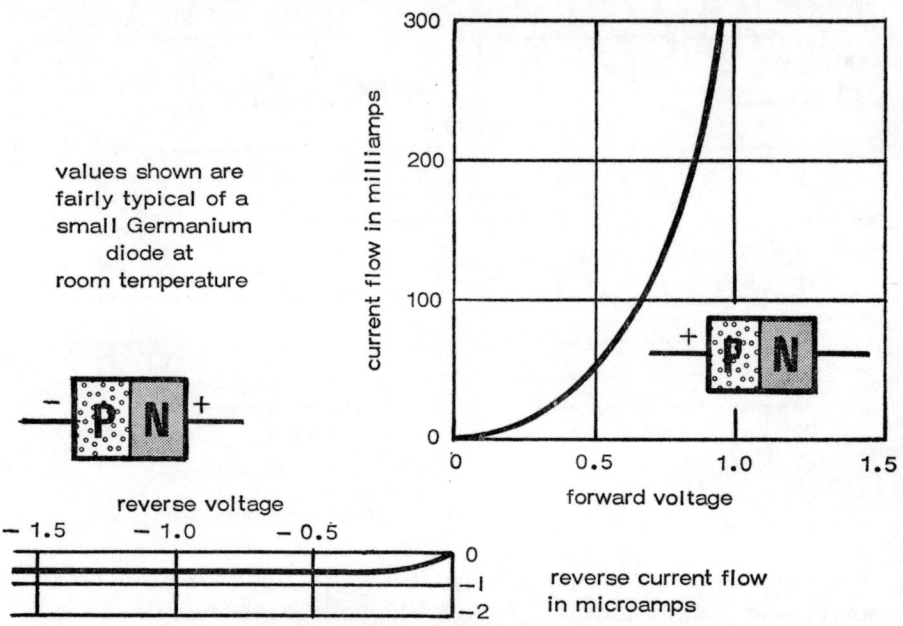

Fig. 14   Characteristic performance of a diode in forward working (conducting) and reverse working (non-conducting).

barrier is neutralized. Typically this will occur at about 0·2 volts with a germanium diode and 0·6 volts with a silicon diode. From then on, increasing voltage increases the current flow, limited only by the specific resistance of a diode material and any other resistance in the circuit. Ultimately, if the current rises too high a figure, the diode would become overheated and burn out.

In the reverse direction the effect of increasing voltage is a very slight increase in leakage current up to the saturation point of the diode. In practice the leakage current can usually be considered constant,

SEMI-CONDUCTOR DIODES AND THEIR CHARACTERISTICS

unless excessive voltage is applied. When this occurs the diode becomes over-saturated and the reverse current rises sharply. If this is not limited to a safe level by another resistance in the circuit the diode will overheat and be destroyed.

One inherent limitation with reverse working (apart from the presence of a leakage current which makes a diode an imperfect rectifier) is that heating effects are present due to power dissipation, and the consequent rise in temperature will automatically result in an increase in the leakage current itself. In the case of a *germanium* diode this can be quite marked, e.g. the leakage current typically doubles with each 9° Centigrade temperature rise. The effect is far less marked with a *silicon* diode, which is often a preferred type for this reason. On the other hand the effect of heating on current flow, when the diode is operating in the *forward* direction, is usually negligible.

Types of Diodes

Two different types of diodes have already been mentioned— *germanium* diodes and *silicon* diodes. Silicon diodes are less sensitive to temperature and thus generally maintain a low reverse current. Working in the forward direction, a higher initial voltage is needed to start the current flowing.

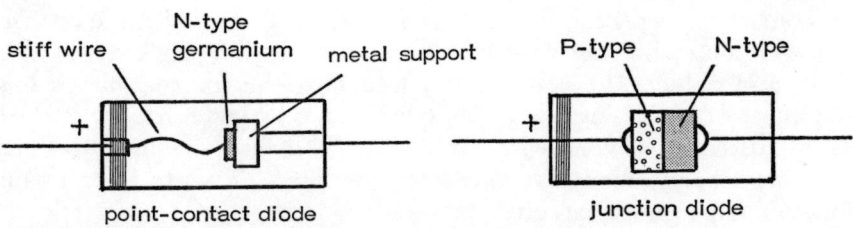

Note: actual length is about one-third of an inch

Fig. 15 Two different types of diode construction.

Diodes may also be described by their construction, e.g. *point-contact* type or *junction* type. With a point-contact diode a small N-type crystal touched by a pointed wire is enclosed in a suitable envelope. It is treated during manufacture to give P-type characteristics to the crystal in the vicinity of the point—Fig. 15. In the case of a *junction* diode, its construction is essentially similar to that of a transistor (*see* Chapter 5), comprising a P-type pellet alloyed or diffused to an N-type crystal slice. Like a transistor it is encapsulated. Point-contact diodes are almost exclusively made from a germanium crystal. Junction-type diodes may be germanium or silicon, but most are made from silicon.

Capacitor Characteristics

Working in the reverse direction, the depletion layer is a non-conducting barrier between the two ends of the diode and so virtually forms a dielectric between two electrodes. In other words, the diode working in this direction is equivalent to a capacitor. Hence it has an ability to pass alternating currents which may be present.

The degree of capacitance effect present depends mainly on the *area* of a junction—the larger the area the greater the capacitance present when working in the reverse direction. In practice, the capacitance effect is not generally significant in the passage of *ac* up to frequencies of about 30–50 Hz, but at very much higher frequencies capacitance effects can be considerable. The only way in which they can be reduced is to reduce the junction area, which is why a point-contact diode is normally preferred for circuits carrying very high frequency currents.

There is another type of diode which uses this capacitance effect in a different way. Beside the junction area, capacitance also depends on the *thickness* of the depletion layer. This, in turn, will vary with the reverse voltage. Increasing the reverse voltage increases the thickness of this layer and reduces the capacitance. Reducing the reverse voltage has the effect of increasing the capacitance. Thus a diode can be worked as a capacitor varying with voltage, such devices being known as variable capacity diodes or *variactors* or *varicaps*. They are often useful for stabilizing tuning circuits.

Zener Diodes

A *Zener diode* is a special form of silicon junction diode with a low and constant *breakdown voltage*. A typical performance curve is shown in Fig. 16 (compare with the performance of a conventional diode, Fig. 14). The sharp break from non-conducting to almost perfect conductance at a particular reverse voltage is called the zener knee. This characteristic makes a zener diode a particularly useful device for producing voltage-stabilizing or current-stabilizing circuits. A zener

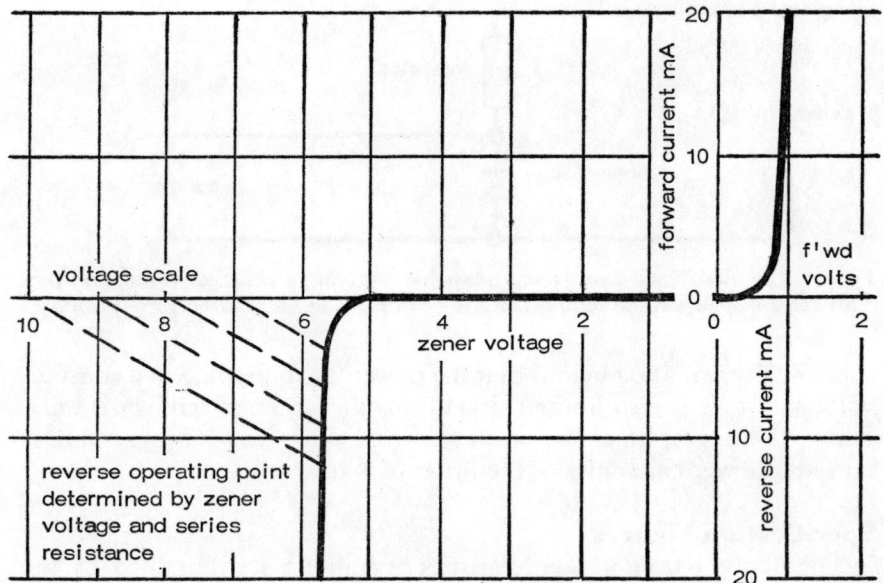

Fig. 16 Working characteristics of a Zener diode.

diode is normally operated at or above breakdown voltage, with sufficient resistance in the circuit to limit the actual current flowing in the circuit to a safe figure for the zener diode used.

Thus in a simple demonstration circuit shown in Fig. 17, a zener diode is connected in a series with a limiting resistor to a supply voltage. This resistor effectively divides the supply voltage into the breakdown voltage across the diode with the remainder dropped across the resistor. The voltage across the zener diode *remains* constant, even if the supply voltage varies. Thus tapping the circuit at this point can provide a constant voltage output. The only thing that happens with a variable supply voltage is that the current flow will rise or fall in proportion. Only if the supply voltage falls so much that the working point of a diode is pulled back past its knee point will the constant voltage output cease.

Only one constant voltage can be tapped from the zener diodes, corresponding to the breakdown voltage or zener voltage as it is generally called. Typically this is of the order of 5 to 6 volts. Being a

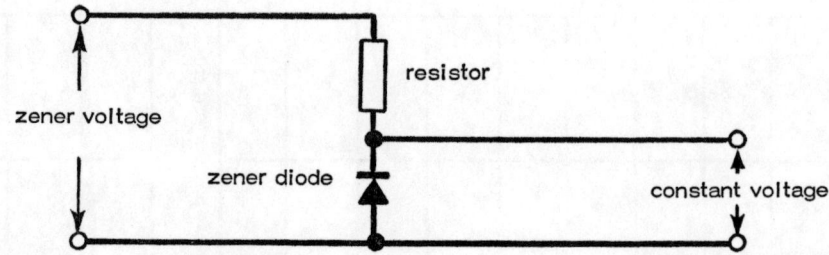

Fig. 17 Simple Zener diode circuit capable of giving a constant voltage output. This can be incorporated in another circuit to provide a stabilized supply voltage.

small device it is also obvious that the power it can provide in a constant voltage circuit is also limited, usually to the order of milliwatts only. Zener diodes are, however, produced with much larger zener voltages and also power capacities of the order of watts.

Specification Figures

The actual working characteristics of a diode are described by the following:

Maximum reverse voltage—with figures normally given for *peak* (or absolute maximum which the diode will tolerate) and *average*. Separate values may also be given for different temperatures, i.e. ambient (usually 20° Centigrade or 25° Centigrade); and at some higher temperatures (usually 60° Centigrade). In the absence of separate temperature ratings ambient temperatures can be inferred.

Maximum forward current—given in milliamps. Again separate figures may be given for *peak* and *average*; and separate values for ambient temperature and 60° Centigrade.

Other data which may be given are:

Ambient temperature range—referring to the maximum and minimum ambient temperatures, within which range the diode will not be harmed.

Maximum junction temperature—usually separate figures for continuous and intermittent operations.

Thermal resistance—in ° Centigrade per millivolt.

CHAPTER 4

THE DIODE DETECTOR

THE *detector* in an AM radio receiver has to work as a rectifier, suppressing half of each cycle of an amplitude-modulated carrier wave signal and converting the current variations in the other half into a direct current fluctuating in the same manner as the original modulation. In simpler terms, it extracts the audio frequency or *af* content of the modulated radio signal in the form of a fluctuating *dc* current.

The semi-conductor diode is a logical choice for achieving this, but its performance is far from perfect. The ideal rectifier would have no forward resistance and infinite backward resistance. Semi-conductor diodes have a low forward resistance and high backward resistance, but the loss of rectification imposed by these characteristics can be recovered by the addition of a *load resistor*, and a *reservoir capacitor* in parallel with it as in Fig. 18.

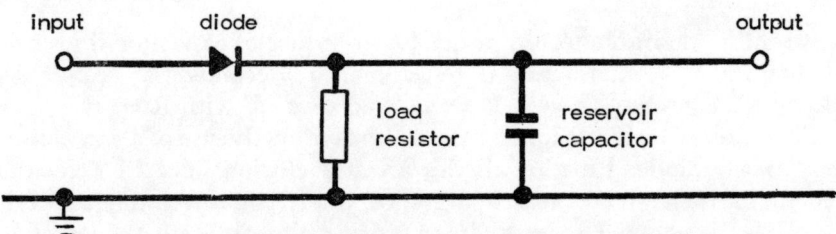

Fig. 18 Basic detector circuit using a load resistor and reservoir capacitor to compensate for the imperfect rectifier characteristics of a diode.

Theoretically, the higher the value of the load resistor the greater the rectified voltage. Its value, however, has also to be chosen to match the input resistance of the next stage. The function of the reservoir capacitor is to maintain the voltage during half cycles when the signal is chopped off. Its value, together with that of the resistor, determines the *time constant* of the circuit. To avoid loss of top frequency and distortion of a signal this has to accommodate the highest modulation frequency likely to be present—say 10,000 Hz for AM radio. This implies a time constant of *not more* than 50 microseconds (the time between a 'peak' and 'trough' of a 10,000 Hz signal), and preferably rather less.

Starting point in deciding suitable component values therefore is first to select R as a suitable match, and then determine the value of C to give a time constant for the circuit of 50 μseconds or slightly less. At the same time the value of the reservoir capacitor must be high enough to maintain a suitably high reservoir action at the lowest radio frequencies present in the signal.

In the case of a straightforward transistor radio receiver the detector is followed by a transistorized audio amplifier stage, the input impedance of which is typically of the order of 1,000 to 2,000 ohms. A suitable value for the load resistor (R) would therefore be 1 k ohm up to possibly 4 k ohm. Corresponding values of the reservoir capacitor to give a CR or time constant of 50 μseconds would be:

$$R = 1k, C = 0.5 \ \mu F$$
$$R = 2k, C = 0.25 \ \mu F$$
$$R = 4k, C = 0.125 \ \mu F$$

It would be desirable to use rather lower values of capacitor than this so that the time consonant is less than 50 μseconds (say 40 or 45 μseconds), e.g. 0.20 μF with R = 2k, and 0.12 μF with R = 4k.

The order of load resistance values also favours the use of a *germanium-point contact* diode. Junction diodes are less efficient as AM detectors because of their greater self capacitance, and quite unsuitable as VHF or UHF detectors. However, in practice quite different values of R and C will work with a germanium diode, but if the time constant is greater than 50 μseconds there will be a loss of top performance and distortion. If much less than about 40 μseconds there will be a loss of detection efficiency at all modulation frequencies.

This simple detector circuit, in fact, allied to a simple tuned circuit (aerial coil), is the basis of a *crystal set*—Fig. 19. In this circuit the load resistor R can be omitted as the high impedance phones will effectively act in its place. It will also work without the reservoir capacitor,

THE DIODE DETECTOR

although with more distortion present. By experimenting with different values of C to adjust the time constant of the circuit, it is possible to establish an optimum value for detection efficiency and/or distortion, but not necessarily both together.

Incidentally, as well as virtually any germanium-point contact diode, almost any *af* transistor will work in this circuit as a detector. The collector lead is ignored and the transistor connected (and worked) as a diode using the base and emitter connections only.

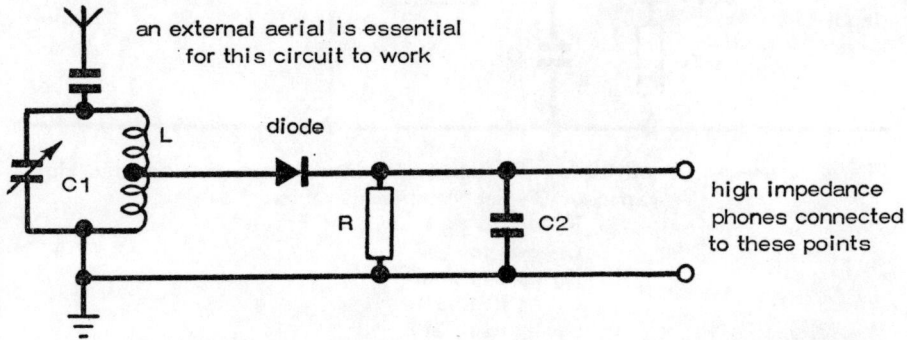

Fig. 19 Coupled to a tuned aerial circuit L and CI, a diode detector completes a 'crystal receiver'. R can be 2·2 k ohm or 4·7 k ohm. C2—0·25 to 8 µF.

A significant feature of this basic detector circuit is that it must be coupled to a high impedance output load to work properly. If a low impedance load is connected directly to a tuned circuit (such as low impedance phones in the case of this circuit) it will have a 'damping' effect which will drastically reduce the efficiency of the tuned circuit and make its response very flat. This applies as a basic rule.

The detector circuit can be improved by removing any unwanted signal content. The diode will inevitably pass a small proportion of *rf* signal which, if allowed to reach the next stage, will be amplified to cause distortion, or even excite self-oscillation. This can be remedied by the adoption of filters—in practice a series resistor and a parallel capacitor.

There will also be a signal present consisting of a *dc* voltage with an average level of the output wave form. Again it is undesirable that this should reach the amplifier stage. It can be removed by inserting a *blocking capacitor* in the output stage.

These additional circuit components are shown in Fig. 20, where R_F and C_F are the resistor and capacitor, respectively, forming the *rf* filter; and C_B is the blocking capacitor. Choice of values is to some extent arbitrary. R_F needs to be fairly small so as not to reduce the

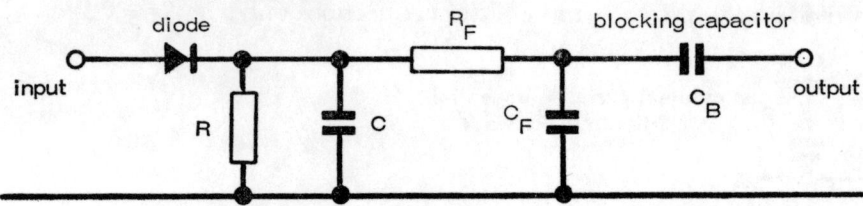

Fig. 20   Detector circuit further developed with filter components and blocking capacitor. Typical component values:
R—2·2 to 3·9 k ohm
C—0·25 to ·5 μF
R_F up to 1 k ohm
C_F—5 to 10 μF
C_B—5 to 10 μF

signal strength unduly—say 1 k ohm or less. The value of C_F is then chosen to give the 'RC' combination the required characteristics to make the filter work, i.e. eliminate all frequencies present above the audible frequency range—say above 15 kHz.

The basic relationship involved is

$$f = \frac{1}{2\pi RC}$$

Written as a solution for RC

$$RC = \frac{1}{2\pi f}$$

taking $f = 15$ kHz
$RC = 10·6 \times 10^{-6}$
Thus if R is 1 k ohm
$C = 10·6 \times 10^{-6}$
or 10 μF, say.

THE DIODE DETECTOR

A similar value is also usually suitable for the blocking capacitor, but almost any value between 5 μF and 10 μF should be suitable.

The complete, practical detector circuit also normally incorporates a volume control or variable resistor R_V, as shown in Fig. 21. The volume control can in fact be placed elsewhere in the receiver circuit—*see later*. The usual value for this potentiometer in a transistorized radio circuit is 5 k ohm. The blocking capacitor can be removed to the output side of the volume control, where it also doubles as a coupling capacitor; or left in the same position as in Fig. 20 if another form of coupling is used. In the former case the value may be adjusted to suit the coupling requirements—*see* Chapter 10. It is also possible to eliminate resistor R

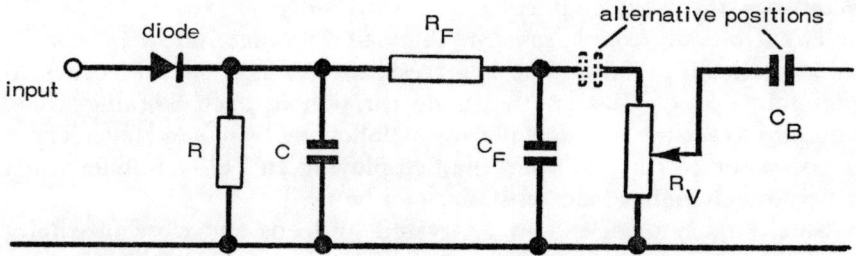

Fig. 21 Circuit of Fig. 20 with volume control added. $R_V = 5$ k ohm. Removal of the blocking capacitor to the output side reduces distortion from the aerial circuit.

in this circuit, choosing a slightly higher value for R_F to compensate (and adjusting the value of C_F to match). This can prove an advantage if the signal strength is weak and it is desirable to increase it by reducing the load resistance in the circuit. It will be an advantage in a crystal set, for example, and in a simple receiver with only moderate amplification following the detector.

Input and Output Resistance

The effective input resistance offered by the tuned circuit is $R/2$, ignoring the forward resistance of the diode which is negligible by comparison. This represents the loading on the tuned circuit, and an opportunity to re-select the value of R if necessary (and at the same time modifying the value of C required as explained earlier).

With the usual 5 k ohm volume control the load resistance at maximum volume will normally be of the order of 2 k ohm to 4 k ohm, depending on the values of R and R_F used.

Detector Performance

Besides being a vital link in the radio circuit, the detector is also responsible for the quality of a signal passed to the next stage. Any distortion or unwanted signal passed on will be amplified. Ideally the detector should act as a perfect linear amplifier. Unfortunately to work at all it has to include non-linear components, which means that *af* currents passed by the detector circuit will not *exactly* follow the modulation of the original *rf* signals. The more marked the non-linear characteristics, the greater the resultant distortion.

The two main trouble spots are the load resistance and stray capacitance. Basically a fairly low load resistance is necessary in order to maintain enough bias on the diode throughout each working cycle, and also to 'match' the detector to the following transistor stage. There is no real answer to this other than employing an FET amplifier when a very much higher load resistance can be used.

Stray capacitances can be generated by leads and close proximity of other current carrying components. To reduce their effect, wires (or printed circuit leads) should be kept as short as possible and the diode should be positioned as far as practicable from leads and other components, although there is no need to carry this to extremes.

FM Detectors

The FM detector has to handle and rectify a quite different type of signal. Instead of the carrier wave being modulated by *amplitude*, it is modulated in *frequency*. Thus for FM reception the detector has to extract the frequency modulation and convert it into a corresponding *af* signal. At the same time it must reject any amplitude modulated signal which may be present. These functions are performed by a *discriminator* and *limiter* respectively. Alternatively, a *ratio detector* circuit can be used which combines functions of both discriminator and limiter. This would seem the simpler solution as the two types of circuit are very similar. However, the *ratio detector* tends to produce slightly more distortion and only has about half the sensitivity of the discriminator, although the latter fact is not necessarily significant.

THE DIODE DETECTOR

There are also two other types of FM detectors which can be used, the *pulse counting detector* based on digital ICs which requires no tuned circuit and has excellent linearity; and a *crystal discriminator* which is manufactured as a sealed unit and requires only an external variable capacitor to set the slope of the detector (i.e. adjust the output voltage so that it swings an equal amount positive and negative).

A basic *discriminator* circuit is shown in Fig. 22. This is, in effect, two AM detectors back to back, fed with rf signal from a tuned circuit $L_1 C_1$ and also receiving half the voltage from a second tune circuit $L_2 C_2$. The same basic circuit will also work as a ratio detector (i.e. as a limiter as well) by reversing one of the diodes and breaking the connection between one side of $L_2 C_2$ and the load resistors and taking the output from these two points.

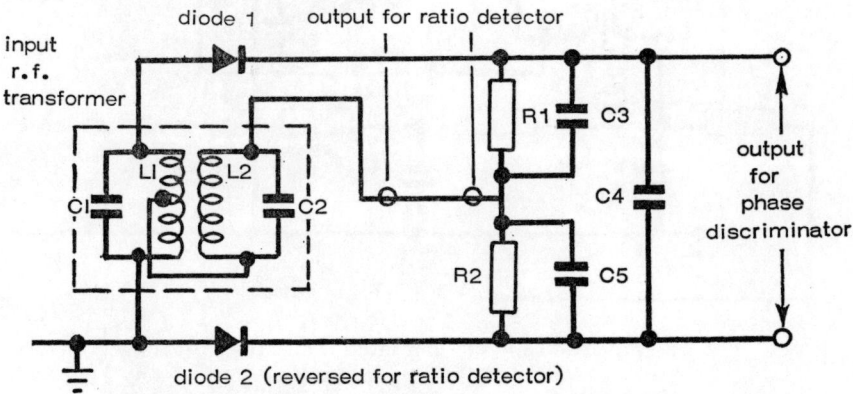

Fig. 22 Basic circuit for a phase discriminator which can also be modified to work as a ratio detector.

R_1 and R_2 are load resistors
C_3 and C_4 are reservoir capacitors
C_5 is a decoupling capacitor (only necessary if the circuit is worked as a ratio detector).

In practice simple circuits are normally used based on an *if* transformer to produce FM-to-AM conversion, operating at the intermediate frequency of a superhet receiver. Voltage developed on one side of the secondary leads the primary voltage, and that on the other side lags by the same phase angle when the circuits are resonant to the unmodulated carrier frequency. This yields equal and opposite voltages passed by the diodes. In the presence of a frequency modulated signal there is a shift in the relative phase of the voltage components, voltage applied to one diode increasing and the other decreasing. The difference between the two, after rectification, is the audio or *af* frequency output of the detector.

A practical ratio detector circuit is shown in Fig. 23, the transformer in this case having three windings. The primary and secondary are

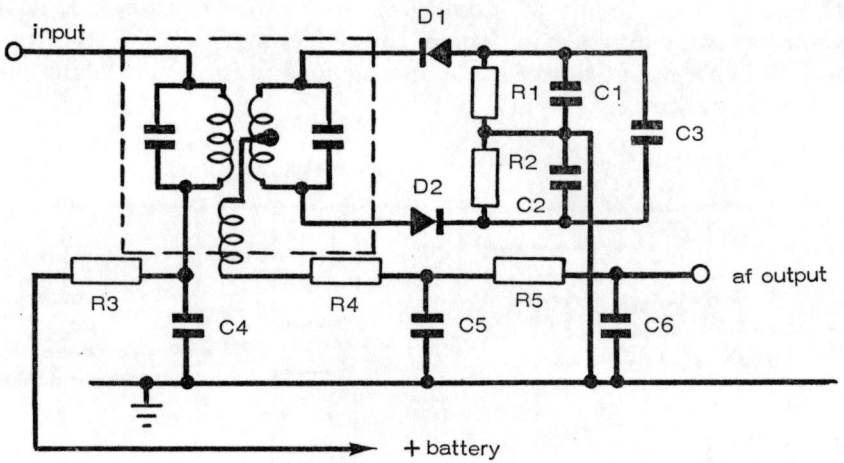

Fig. 23 Practical ratio detector circuit. Component values:
R1 and R2—4·7 k ohm
R3—2·2 k ohm
R4 and R5—68 ohm
C1 and C2—330 pF
C3—8 or 10 µF
C4—0·01 µF
C5—330 pF
C6—0·001 µF
Diodes—point contact type

tuned to the *rf* frequency. The close coupled tertiary injects a voltage into the secondary in series with the voltage across each half of the secondary. It works essentially as a discriminator, with the reversed diode acting as a limiter.

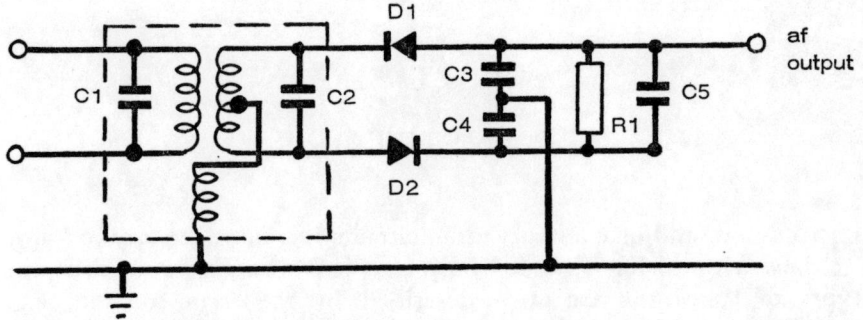

Fig. 24 Simplified practical ratio detector circuit. Component values:
  R_1—20 k ohm
  C_1—180 pF
  C_2—56 pF
  C_3 and C_4—220 or 330 pF
  C_5—10 µF
  Diodes—AA119

A more simplified ratio detector circuit is shown in Fig. 24, capable of giving just as good results.

CHAPTER 5

TRANSISTORS

To design and/or construct radio circuits it is not necessary to know how a transistor is made—only how it works. However, different types of transistors are often described by their construction, e.g. junction type, silicon planar type or epitaxial. The following brief descriptions of the different transistor manufacturing processes can therefore be helpful in giving meaning to these descriptions. Mention of point contact transistors is omitted since this was the original form of construction, now obsolete.

The working characteristics of a transistor depend on the behaviour of a P-type (positive type) and an N-type (negative type) semi-conductor material in a three-element combination. Such combinations yield either a P-N-P or N-P-N device.

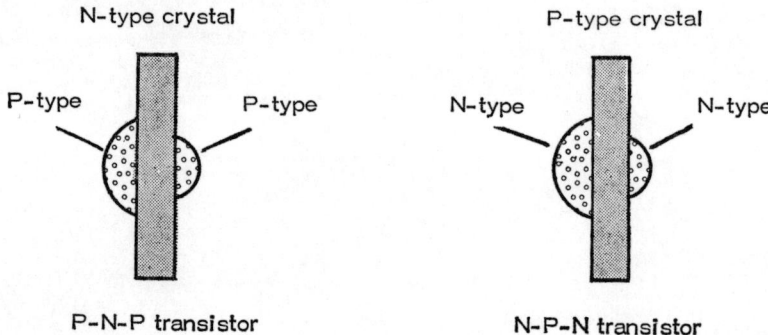

Fig. 25 Construction of P-N-P and N-P-N transistors shown in elementary diagrammatic form.

Starting point in the manufacture of a transistor is the production of a crystal of either P-type or N-type *germanium* or *silicon*, which is cut into slices. This is the *base* element of the transistor. The simplest method of construction is then to fuse a pellet of opposite type material to each of the slices, i.e. two pellets of P-type material to an N-type slice to make a P-N-P transistor; or two pellets of N-type materials to a P-type slice to make an N-P-N transistor—Fig. 25.

These pellets form a *collector* and *emitter* of the transistor. Wires are soldered to each of the three elements and the complete device is then encapsulated—Fig. 26. Transistors made in this fashion are known as *alloy-junction* types, the method being widely used for production of *germanium* transistors.

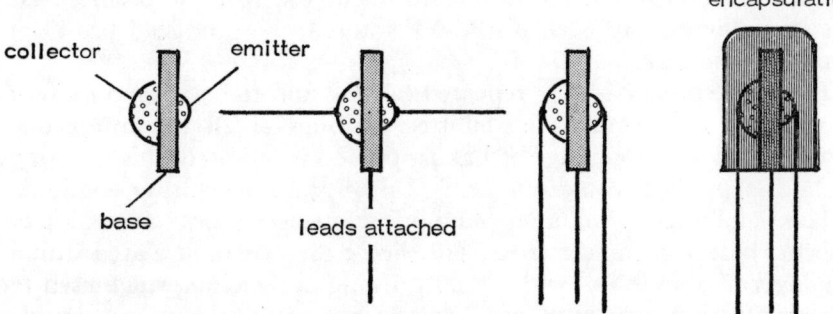

Fig. 26   Basic stages in the manufacture of a junction-type transistor.

A slightly different method is used to produce a junction transistor by *alloy diffusion*. Suppose a P-N-P transistor is required. A layer of N-type material is deposited on one side of a P-type slice. On this are deposited two pellets. One contains N-type additives only and the other both P-type and N-type additives. This assembly is now heated in an oven causing the two pellet materials to diffuse through the N-type layer into the P-type slice. In this case the original slice becomes the collector, the diffused N-type additive the base and the diffused N-type and P-type additive the emitter.

In the case of a *silicon* transistor a different method of construction is now normally used, known as a *planar* process. A slice of N-type silicon (which eventually becomes the *collector*) is heated to form a thin layer of silicon oxide. It is then given a thin coating of photo-sensitive material resistant to etching, known as *photoresist*.

This coated slice is used for the manufacture of several transistors. It is first covered by a photographic negative of the pattern of the base positions of the number of transistors to be made and exposed to ultra-violet light. This prints the base patterns on to the slice. The slice is then washed, which removes all the unexposed photoresist, leaving a print of several small windows in the photoresist through which the silicon oxide is exposed. The slice is then etched to remove the silicon oxide appearing in the windows, exposing silicon.

After cleaning, the slice is treated in an oven with some P-type doping additive. This additive evaporates and some settles on the windows. Further heating in a second oven in an atmosphere of oxygen diffuses the P-type dope into the silicon at the window position. At this stage the base of each of the transistors to be produced has been formed in the slice.

The whole process is then repeated to print and etch a second pattern of windows representing the emitter positions of all the transistors. After etching and cleaning the baking process is repeated, this resulting in the N-type doping additive being diffused into the emitter windows.

This is followed by another photoprinting stage to produce windows over the base and emitter areas. The slice is then vacuum plated with a fine layer of aluminium and a final printing and etching stage used to remove aluminium from all areas except where it is required to provide contacts with the bases and emitters. At these points the aluminium is lightly alloyed to the bases and emitters.

The oxide layer is now removed from the collector side of the original slice and the thickness of the slice reduced as required. After testing each transistor so formed, the slice is then cut into separate pieces each containing one transistor. Faulty transistors marked at the testing stage are rejected. Every good transistor is then bonded, collector side down, to a gold plated stem. These wires are bonded to the aluminium connections to the emitter and base. Finally the transistors are encapsulated.

The process is considerably more complicated than the manufacture of alloy-junction or alloy-diffused transistors, but a single slice may yield from 200 to 7,000 or more transistors, depending on type. Stages of manufacture are illustrated in Fig. 27.

A somewhat modified process is used in the production of *epitaxial silicon planar transistors*. Basically the original slice is more heavily doped, with the result that it is covered with a very thin layer of highly doped silicon with the same crystal orientation as the original material, this being known as the *epitaxial* layer. In the final product the transistor

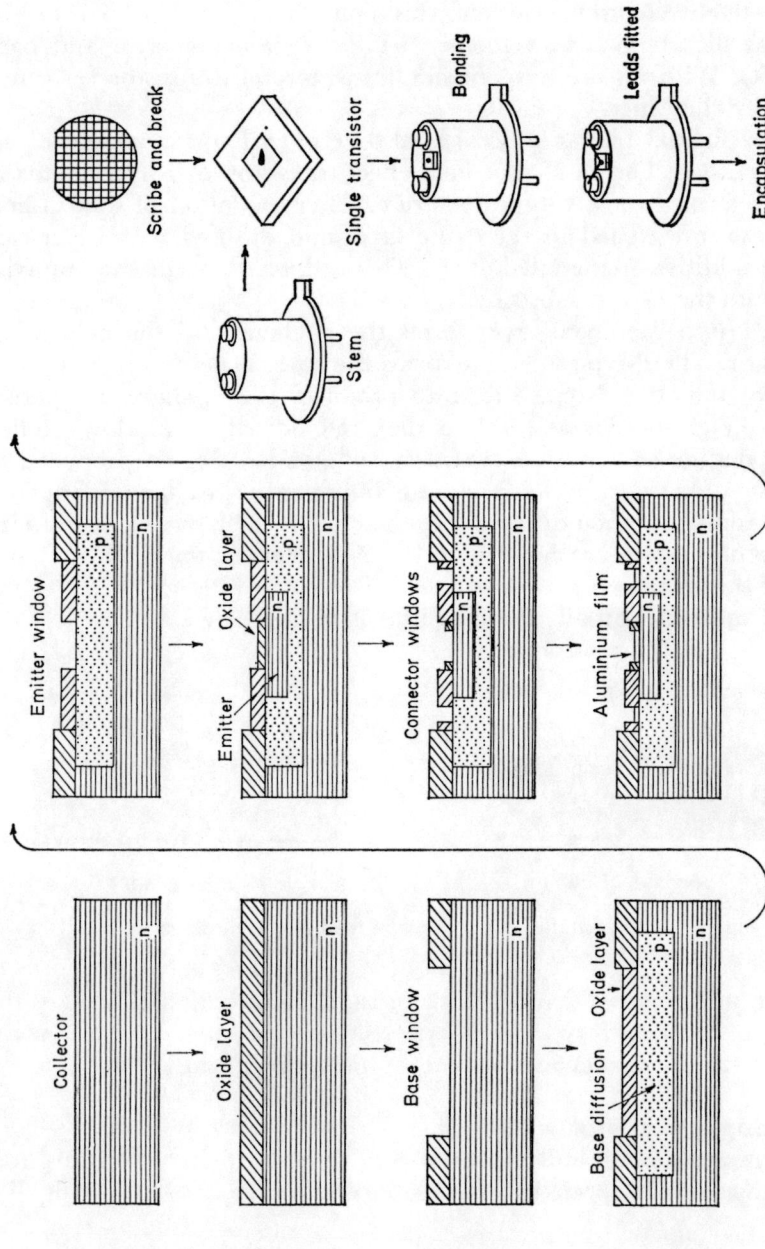

Fig. 27 Stages in the construction of silicon planar transistor (diagrams by Mullard).

is effectively contained within this epitaxial layer and not in the original slice (which now merely acts as a stiffening layer and back contact). With an ordinary planar transistor the transistor is formed in the original slice.

The epitaxial planar process is also employed to manufacture *field effect transistors*. For an N-type field effect transistor an N-type epitaxial layer is formed on a P-type substrate. After oxidation of the surface, 'windows' are etched in the oxide layer and diffused with super-rich P-type additive (designated P+). This diffuses through the epitaxial layer into the P-type substrate.

The N-type epitaxial layer forms the N-channel of the field effect transistor. The P-type substrate forms the gate. Penetration of the P+ additive into the P-type substrate produces the N-channel isolating region—Fig. 28. The assembly is then re-oxidized, a 'window' etched corresponding to a second area and another P+ region produced to form the upper part of the P-gate surrounding the N-channel. Super N (N+) additive is then diffused into the ends of the N-channel. Aluminium contact areas are bonded to the N+ regions, to which are connected the *source* and *drain* lead wires. The *gate* lead is connected to the P-type substrate. Finally the transistor is encapsulated.

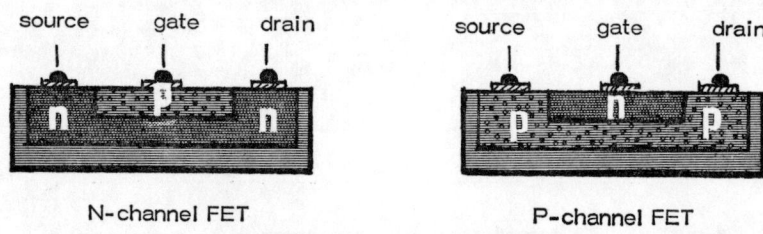

Fig. 28 Elemental construction of a field effect transistor or FET

With a P-channel field effect transistor, construction is exactly the same, substituting P-type for N-type elements, and vice versa. *See also* Chapter 11 for further information on field effect transistors.

Selecting Transistors

There are now literally thousands of different designs of transistors made by various specialist manufacturers. In the case of circuit diagrams, or component catalogues, they are identified only by a type number, which is the manufacturer's coding. This normally consists of letters and numbers, the letters usually identifying the manufacturer or general class of the device with the associated numbers indicating the particular design or development. Thus 'O' indicates a semiconductor device manufactured by Mullard, and an appended 'C' that the device is a transistor. Mullard transistors are thus coded 'OC' followed by a serial number identifying the particular type. This does not help at all unless such a manufacturer's specific type is quoted on a circuit diagram, which it usually is. The reason why a type number should be quoted for a specific circuit design is quite simple. Possible variations in the design performance of a transistor are virtually infinite, so even if different manufacturers produce transistors for a similar purpose, the actual characteristics of different makes can vary quite widely. This can affect choice of matching components which govern the proper working of the circuits.

In certain cases, particularly in the field of more or less 'standard' radio circuits, transistor types from different manufacturers may have sufficiently close characteristics to be regarded as equivalents or nearequivalents. This implies they could be substituted for a specified type in a particular circuit. In many more cases—usually with very elementary circuits—almost any type of transistor of the same basic type, e.g. germanium or silicon, or *functional group* and polarity, may produce similar working results.

It is particularly useful to the designer/experimenter to have access to information both on *transistor equivalents* and *functional groupings*. The former information is well catered for in separate publications on the subject (complete books on their own!). Information on functional grouping is much harder to come by. Manufacturers' data sheets are not readily available to the amateur designer/constructor and a comprehensive coverage would need a complete filing system to deal with. Retail catalogues are not usually particularly helpful, either, as they normally list types stocked merely by type number, although there are exceptions. The latter can provide a basis of compiling a list of transistor types readily available by functional grouping, which can be further expanded by reference to transistor equivalent data.

There is a welcome trend to provide more information about transistor types and manufacturers' coding, an example being the

Pro-Electron coding (now adopted for all new Mullard transistors, for example). This consists of a two letter code followed by a serial number. Restricting the code used only to diodes and transistors:

the first letter indicates the semi-conductor material, i.e.
   A = germanium
   B = silicon
the second letter indicates the general function of the device, i.e.
   A—detector diode or mixer diode
   B—variable capacitance diodes
   C—*af* transistor (excluding power types)
   D—*af* power transistor
   F—*rf* transistor (excluding power types)
   L—*rf* power transistor
   S—switching transistor (excluding power types)
   V—power type switching transistor
   Y—rectifier diode
   Z—zener diode

The *serial number* following then consists of three figures if the device is intended for consumer applications, e.g. radio receivers, audio amplifiers, television receivers, etc. A serial number consisting of a single letter followed by two figures indicates the device is intended for industrial or specialized applications.

For the average amateur radio enthusiast, one of the most useful *functional groupings* published is that of the Electrovalue Catalogue, albeit listing only a limited number of individual types in each group. This covers the following:

Germanium Transistors
   (i) small, medium currents switching services.
   (ii) medium current switching, low power output.
   (iii) small, medium current amplifiers.
   (iv) *af* amplifiers, low power output.
   (v) complementary pairs.
   (vi) high power output (power transistors).

Silicon Transistors
   (i) *af* amplifiers, small signal, general purpose.
   (ii) *af* amplifiers, low level, low noise.

TRANSISTORS

(iii) small signal amplifiers.
(iv) *rf* amplifiers and oscillators.
(v) medium current switching, low power output.
(vi) high frequency, medium powers.
(vii) general purpose switching.
(ix) power transistors.

Any reference to 'power' grouping is largely arbitrary since there is no universal agreement on the range of power levels (referring to the maximum power rating of the particular transistor). Thus *low power* may generally be taken to cover 100–250 mW, but such a grouping may include transistors with power ratings up to 1 watt. Similarly *medium power* implies a possible power range of 250 mW to 1 W (but may extend up to 5 watts). Any transistor with a power rating of greater than 5 watts is classified as a power transistor.

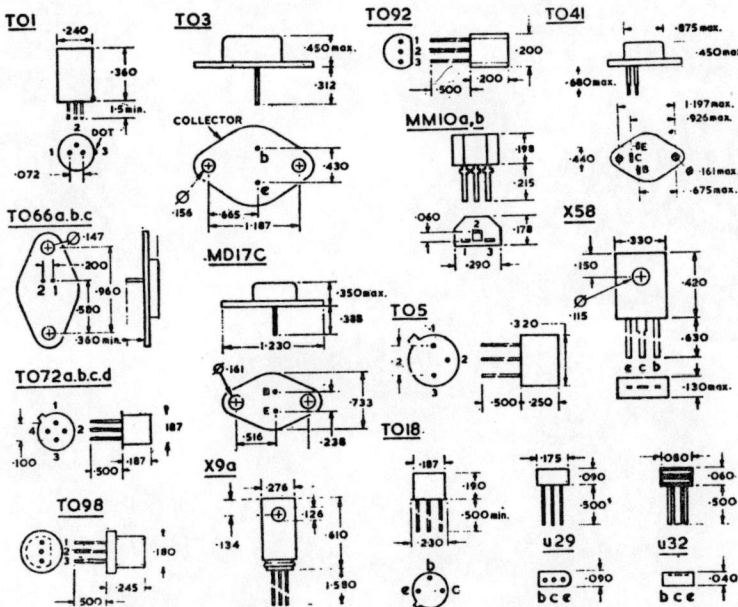

Fig. 29 Representative transistor outline shapes with dimensions and lead identification. Dimensions are in inches.

It is also useful to be able to identify transistors by their size and shape, and *essential* to be able to identify base, collector and emitter leads correctly. Representative outline shapes, with dimensions in inches, are given in Fig. 29, the coding given being more or less universal but not always quoted in transistor specifications and still less so in catalogues.

CHAPTER 6

UNDERSTANDING TRANSISTOR CHARACTERISTICS

PHYSICALLY a transistor is a three-element device (although it is more correctly known as a *bipolar* device). It consists of a combination of P-N-P or N-P-N substances. The first two letters in these designations indicate the respective polarities of the voltages applied to the emitter and collector, in normal operations, i.e.

P-N-P—*emitter* is *positive* with respect to both collector and base; and the *collector* is made *negative* with respect to both the emitter and base.

N-P-N—*emitter* is *negative* with respect to both collector and base; and the collector is made *positive* with respect to both the emitter and base.

More simply—

P-*N*-P—the *collector* must be *N*egative.
N-*P*-N—the *collector* must be *P*ositive.

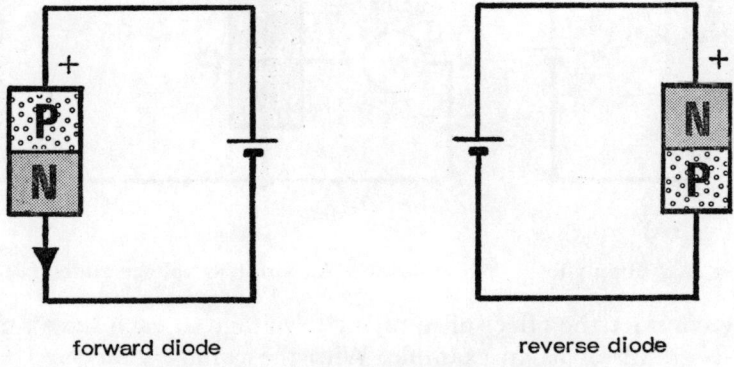

forward diode   reverse diode

Fig. 30   Forward and reverse diode circuits.

These correct polarities are obtained by applying *bias* of the right polarity to each 'side' of the transistor. To get a clear picture of what these 'sides' are, consider the characteristics of a diode (Chapter 4). Depending on the polarity applied to it, it works on the reversed bias mode with no current flow, or in the forward direction with the characteristics of a conductor allowing current to flow—Fig. 30.

Now consider a pair of diodes joined with a common centre to make a three-element device, either N-P-N or P-N-P—Fig. 31. (The diagram shows a N-P-N combination.) This is basically what a transistor is, but with the centre layer made very thin.

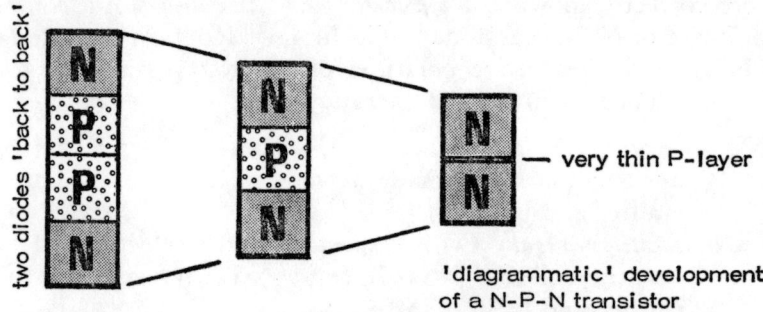

Fig. 31 Two 'diodes' joined 'back-to-back' to make a transistor.

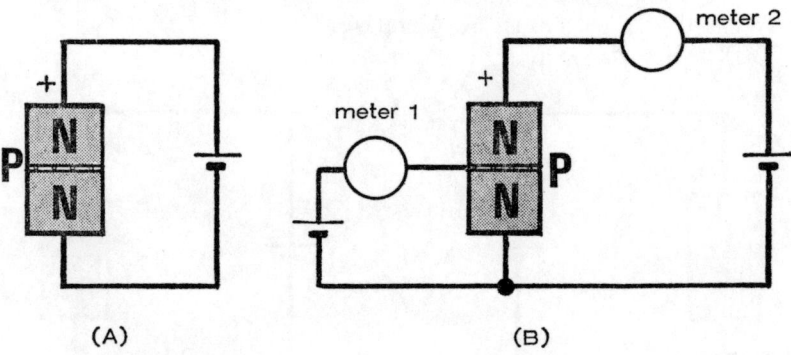

Fig. 32 Supply for N-P-N transistor (A) with bias voltage added (B).

Now consider the effect of applying a voltage to each 'half', taking a N-P-N transistor as an example. With the voltage connected to the 'top and bottom' as in Fig. 32 A, the 'top' half is in the condition of a

UNDERSTANDING TRANSISTOR CHARACTERISTICS

reversed-biased diode, so no current will flow (other than a small leakage current). This voltage so applied across the whole of the transistor is known as the *supply* voltage.

Applying a second source of *emf* or *bias voltage* to the middle layer and the 'bottom' as in Fig. 32 (B) will immediately alter the situation. The bottom 'half' is now working as a forward diode. This will excite a flow of electrons into the upper part (the reason for making the middle layer very thin is to make this flow easy to establish), so that the whole of the transistor becomes conductive as far as the *supply* is concerned. In other words application of the correct bias produces a small current flowing through the left hand side of the circuit, with a very much larger current flowing through the right hand part of the circuit. These circuit currents can readily be measured by meters at meter 1 and meter 2 positions, respectively. Fig. 33 shows this mode of working for both a N-P-N and P-N-P transistor.

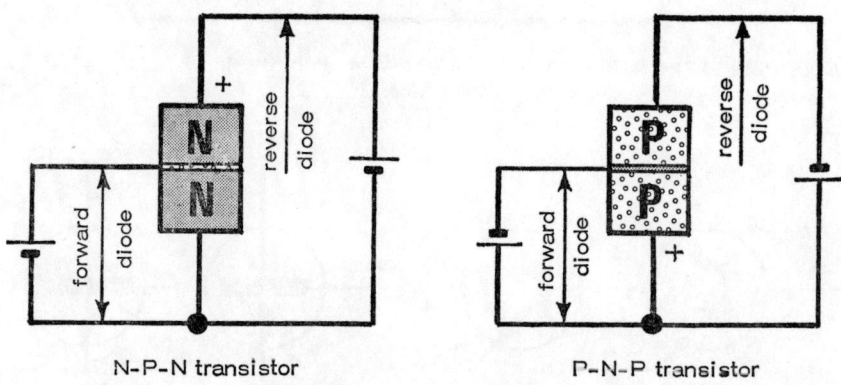

Fig. 33   Supply and bias connections to N-P-N and P-N-P transistors.

Still visualizing the transistor in block form, with the supply and bias connected as in Fig. 34, the basic working conditions are:

(i) The bottom 'half' is a source of free electrons when correct bias is applied, hence the name *emitter*.

(ii) The top 'half' readily accepts or collects these electrons when the transistor is working, and is known as the *collector*.

(iii) The middle layer is a point of application of the bias. It could have been called the 'bias', but that description is reserved for voltages. Alternatively it could have been called the grid, for it works like the control grid of a triode valve. However, for no particular reason which is clear, it is called the *base*.

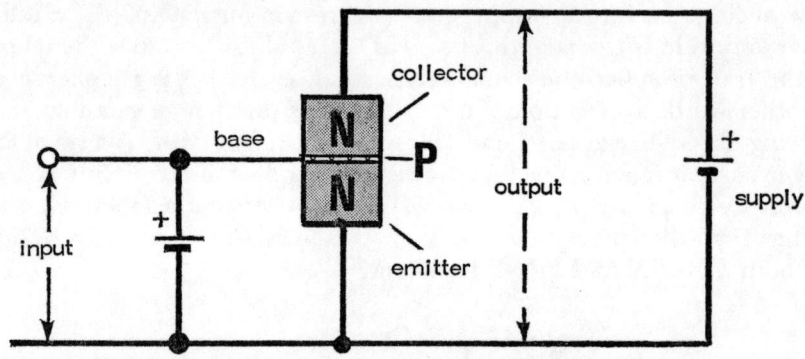

Fig. 34 Input and output circuits of a transistor. Polarity shown is for an N-P-N transistor.

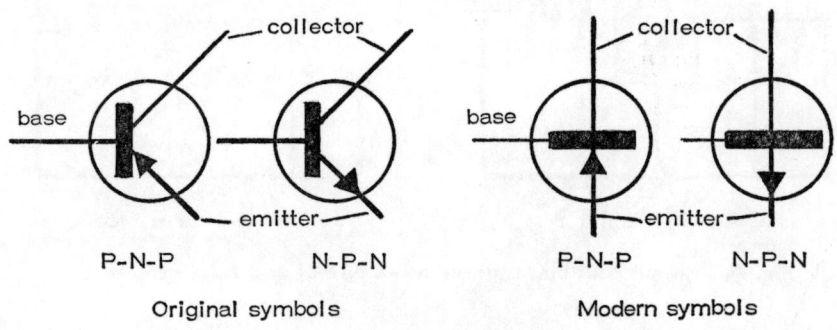

Fig. 35 Transistor symbols.

Thus the three elements of a transistor are called the emitter, collector and base, usually designated by the lower case letters e, c and b respectively. They are represented symbolically as shown in Fig. 35 (sometimes the enclosing circle is omitted):

UNDERSTANDING TRANSISTOR CHARACTERISTICS

(i) the base by a thick black line.
(ii) the collector by a thin line.
(iii) the emitter by a thin line with an arrow. The arrowhead points in the direction of *positive* current flow, so it is the opposite way round for P-N-P and N-P-N transistors.

Basic connections, with appropriate polarity, for P-N-P and N-P-N transistors, are shown in Fig. 36.

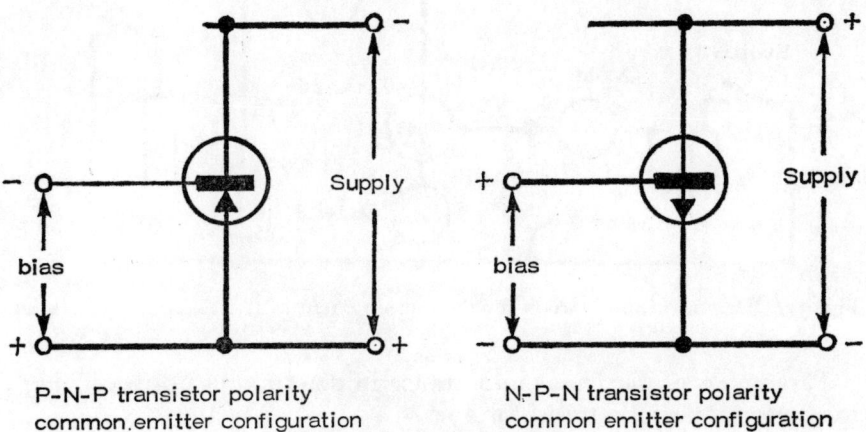

P-N-P transistor polarity
common emitter configuration

N-P-N transistor polarity
common emitter configuration

Fig. 36 Polarity of bias and supply for working of P-N-P and N-P-N transistors in common emitter configuration.

We can now draw a 'demonstration' circuit for a transistor in which its working characteristics can be measured and defined—Fig. 37. Provision is included to vary both the supply voltage and bias voltage and meters are included in the circuit where shown to read:

Meter 1 (microammeter)—base current or I_B
Meter 2 (voltmeter)—base voltage or V_B
Meter 3 (milliammeter)—emitter current or I_E
Meter 4 (milliammeter)—collector current or I_C
Meter 5 (voltmeter)—collector voltage or V_C

With switch 2 closed and switch 1 open the collector/base of the transistor acts as a reverse (diode) junction. The only current that can flow through the reverse junction is a very small leakage current. There will, in fact, be no current flow unless the bias circuit is completed, but this only requires a very small bias voltage or bias current, virtually equivalent to $V_B = 0$ or $I_B = 0$ with switch 1 closed. For most practical purposes this leakage current can be ignored.

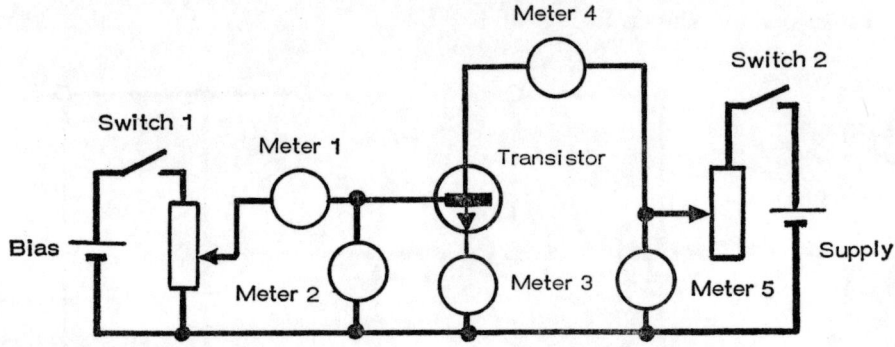

Fig. 37 'Demonstration' circuit for the measurement of transistor characteristics.

Parameters of particular importance in determining (or specifying), the performance of a transistor are:

(i) *Output Characteristics*—or how collector current varies with bias and collector voltage (supply voltage). Bias is usually expressed in terms of base current.

(ii) *Input Characteristics*—or how base current varies with base voltage.

(iii) *Transfer Characteristics*—or the relationship between collector current and base voltage.

(iv) *Current Amplification*—or the ratio of collector current to base current or signal current gained.

(v) *Ratio of Collector to Emitter Current*—This was originally known as the alpha (α) of a transistor, now designated *hfb*.

The *output characteristics* (I_C/V_C) of a transistor can be measured in terms of how the collector current (I_C) varies with collector voltage (V_C) for different values of base current (I_B). Typical curves are shown in Fig. 38 where it will seem that above the transistor 'knee' point, collector

UNDERSTANDING TRANSISTOR CHARACTERISTICS

current is fairly constant with increasing collector voltage for all values of bias (base current). It also indicates that the *output resistance* of a transistor is quite low.

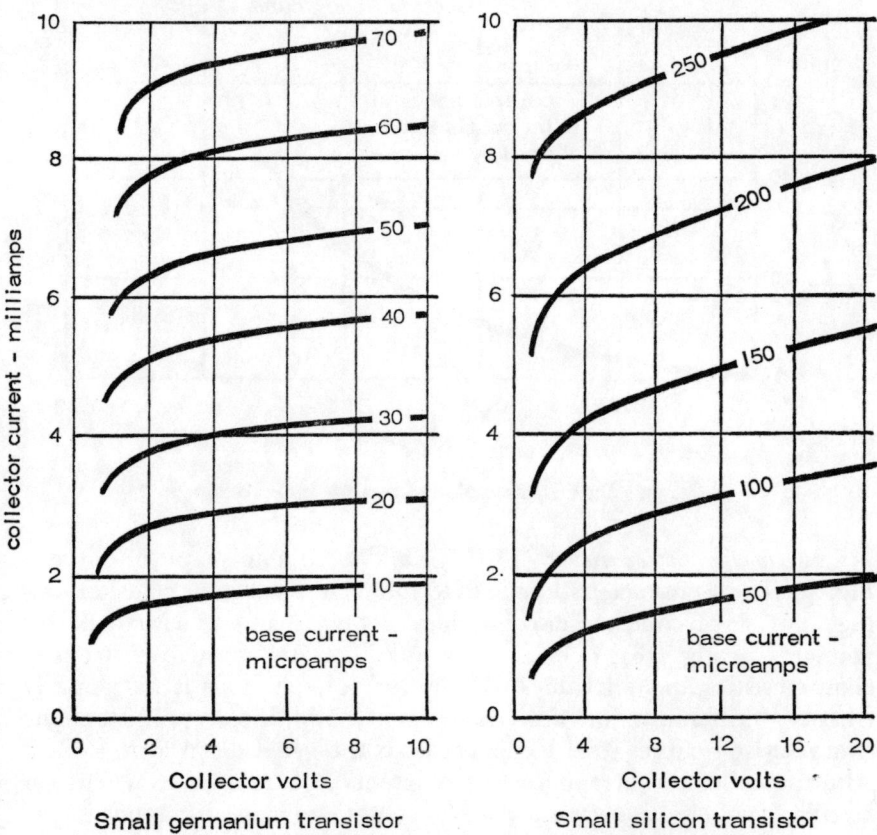

Fig. 38 Characteristic curves for small transistors.

The *input characteristic* (I_B/N_B) can be plotted by measuring base current (I_B) against base voltage (V_B) for a constant collector voltage. In this case a typical form of the curve is as shown in Fig. 39. The *input resistance* at any point in this case is represented by the gradient of the curve or the ratio of V_B to I_B.

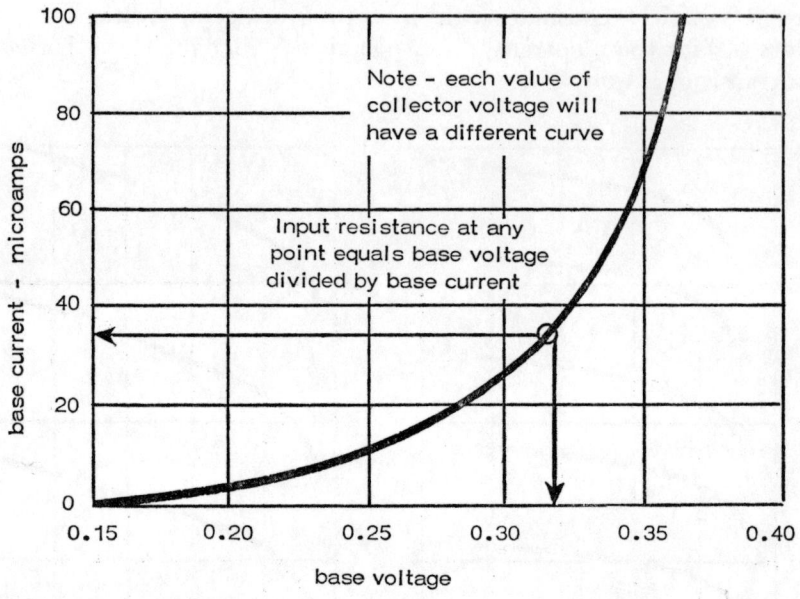

Fig. 39   Base current plotted against base voltage.

The *transfer characteristics* (I_C/V_B) yield a similar shape of curve—Fig. 40. These characteristics are determined at a constant collector voltage but since collector current does not vary much with collector voltage (*see* Fig. 38), one curve is usually representative of transfer characteristics and all values of collector voltage. This is particularly true of germanium junction transistors. With silicon transistors the characteristic curves tend to depart from the generalized form shown. Also the collector current (leakage current) and zero base current (or zero base voltage) is very much smaller. But the main difference in the characteristics is that a higher base voltage (bias) is needed to cause an appreciable amount of collector current to flow in the case of a silicon transistor, i.e. to operate the transistor above the 'knee' point. This bias is about 0·6–0·7 volts for a silicon transistor but only 0·1–0·2 volts for a germanium transistor.

The *current amplification factor* (I_C/I_B) can be derived by plotting collector current against base current. Typically this gives a straight line relationship—Fig. 41. This was originally called the beta(β) of a transistor but is now designated *hf* or *hfe*. Actual values may range from as low as 10 up to several hundreds, depending on the type of transistor, also the slope of the I_C/I_B curve is not always constant, i.e., the value of *hf* may have a spread of perhaps 50 per cent on either side of a nominal value.

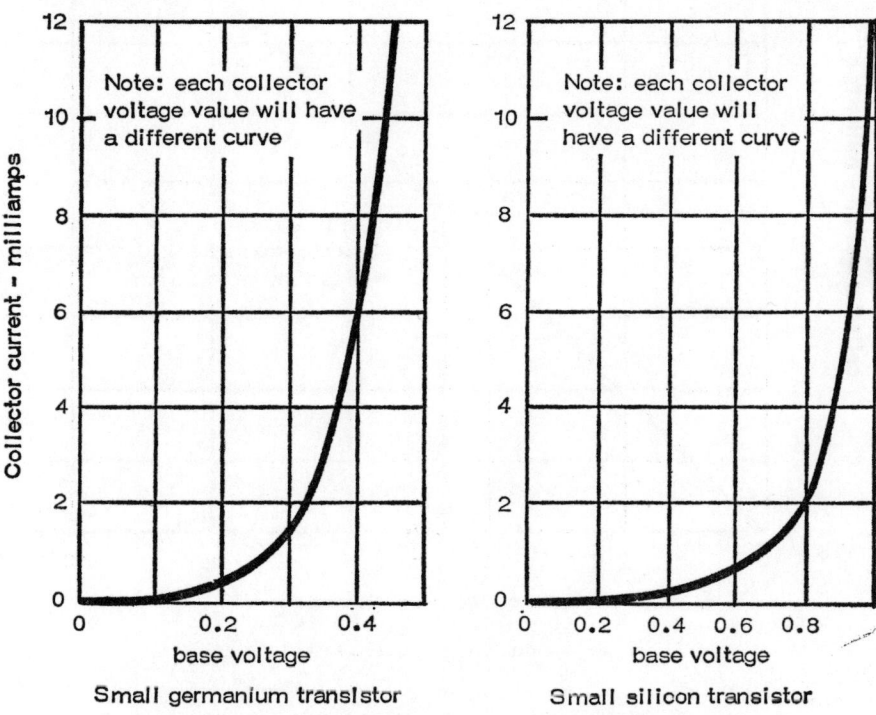

Fig. 40 Collector currents plotted against base voltages.

The *ratio of collector and emitter currents* (I_C/I_E) is not of particular significance. Emitter current must equal $I_C + I_B$, but since I_B is small in comparison with I_C, emitter current (I_E) is roughly equal to I_C. The smaller the base current the closer the collector current comes to being equal to the emitter current, and thus the closer the ratio of *hfb* comes to unity. In practice, typical values of *hfb* lie between 0·92 and 0·98.

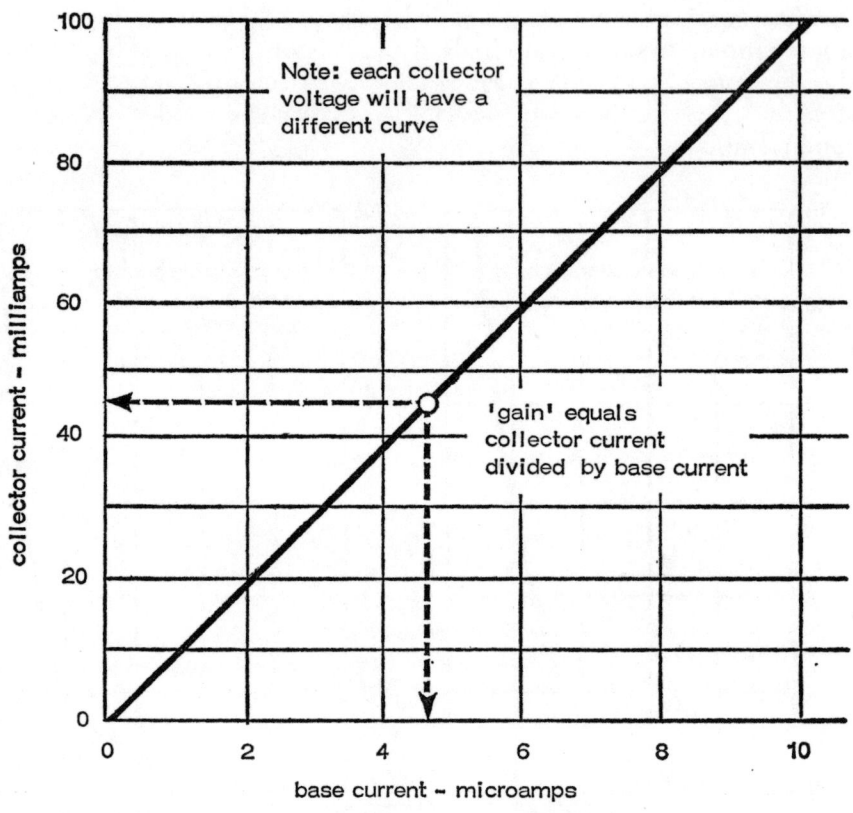

Fig. 41 Collector current plotted against base current.

Ratings

Individual types of transistors are also given *ratings*, representing maximum values the transistor can handle in a circuit. These are:

Maximum collector—base voltage with emitter open circuit, designated Vcbo max.
Maximum collector—emitter voltage, designated Vce max.
Maximum base—emitter reverse voltage or bias, designated Veb max.
Maximum collector current, designated Ic max.

Maximum collector—base current with emitter open circuit, designated Icbo max.

Maximum total power dissipation, designated Pt max.

Circuit Considerations

The three elements of a transistor represent separate resistance in a simple direct circuit as shown in Fig. 42. The *input* circuit embraces the base resistance (r_b) and the emitter resistance (r_e). The *output* circuit embraces the collector resistance (r_c) and any load resistance (R_L), and the emitter resistance (r_e). In other words the emitter resistance is common to both input and output circuits.

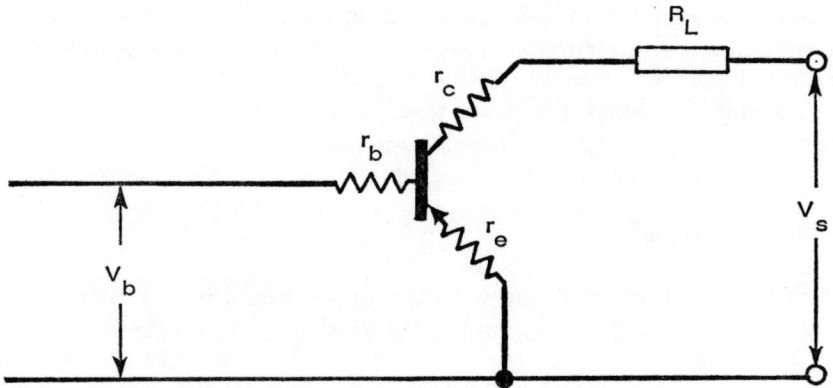

Fig. 42 Resistive elements of a transistor.

The *input resistance* to a simple direct voltage is represented by the base resistance (r_b) and the emitter resistance (r_e) in series. The emitter resistance is small compared with the base resistance—typically 10 ohms as compared with 1,000–3,000 ohms for a small transistor. Hence the base resistance is more or less equal to r_b, with the following relationship applying

$$r_b = \frac{V_B}{I_B}$$

However, when the transistor is actually working the presence of an input excites an amplified current in the output, this current being equal to $I_B \times hfe$. The effective emitter resistance is thus modified by feedback and becomes $r_e + r_e \times hfe$. Thus the total effective input resistance is

$$r_b + r_e (1 + hfe)$$

The effect of r_e can be ignored as it is working as a reversed junction where its resistance is high enough not to have a shunting effect on the input circuit.

This relationship only holds for as long as the output load is zero. Under realistic working conditions with a positive load the collector current (I_c) is *less* than $I_B \times hfe$ and so the feedback is less. In other words the effective input resistance *decreases* with increasing loads. The amount of which this resistance is reduced is directly related to the conductance in the circuit, i.e. the intrinsic inductance represented by $1/r_e$ and the load inductance represented by $1/R_L$.

To simplify calculation both the feedback and conductance effects can be ignored, using the original equation

$$\text{input resistance} = r_b = \frac{V_B}{I_B}$$

Both the current gain and voltage gain are also affected by feedback, but again ignoring this for simple calculation

$$\text{current gain} = hfe$$
$$\text{voltage gain} = \frac{hfe \times R_L}{r_b}$$

These formulas will exaggerate both current gain and voltage gain, but not to any great extent under normal working conditions. Certainly the difference will usually be within the spread of characteristics of individual transistors of the same type.

Bias Circuits

The basic circuit of Fig. 37 derives base bias from a separate battery. In any practical circuit (other than a test or demonstration circuit), base bias voltage is normally taken from the collector supply. There are various ways in which this can be done, the simplest being to 'tap'

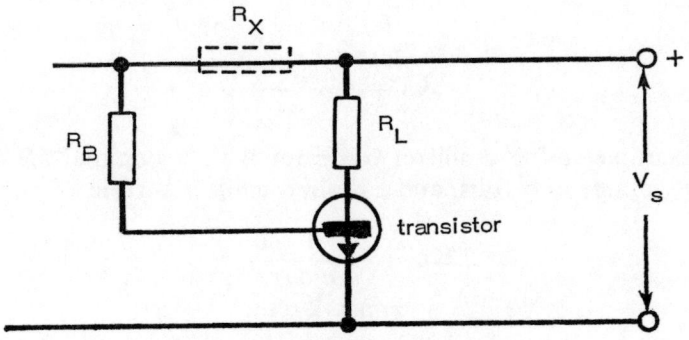

Fig. 43   Simple current biasing.

the collector supply via a bias resistor (R_B) as in Fig. 43. R_L is the collector load resistance. The value of the bias resistor can be calculated from the supply voltage (V_S), as well as the bias required.

$$R_B = \frac{V_S - V_{BE}}{\text{bias required or base current } (I_B)}$$

V_{BE} is the base-to-emitter voltage. In practical circuits this is usually of the same order as the 'starting' voltage of a transistor, e.g. 0·2 volts for a germanium transistor and 0·7 volts for a silicon transistor with a possible spread of 0·05 volts on either side.

Knowing the bias current required (from a transistor's characteristics), which will usually be in *microamps* for small transistors, and using numerical values for the voltages, the value of the bias resistor is given in *megohms* (M ohms). Not knowing the transistor characteristics, a typical value could be used, e.g. 30μA for a small transistor. Calculation can be simplified further by ignoring V_{BE}.

More accurately, the bias resistance should be calculated from the desired value of the emitter current (I_E)

$$I_E = \left(\frac{V_S - V_{BE}}{R_B}\right)(1 + hfe)$$

$$\text{where } R_B = \frac{(V_S - V_{BE})(1 + hfe)}{I_e}$$

Again V_{BE} can be ignored to simplify calculation, when

$$R_B = \frac{V_s \times hfe}{I_e}$$

For example, using a silicon transistor with a nominal *hfe* of 100, a supply voltage of 9 volts, and a desired emitter current of 1 mA,

$$R_B = \frac{9 \times 100}{0.001}$$
$$= 900 \text{ k ohm}$$

The value calculated from the 'complete' formula will be $\frac{8.3 \times 101}{0.001}$

$$= 838 \text{ k ohm}.$$

Nearest preferred value (i.e. actual resistance values obtainable) would be 820 k ohms in either case. In other words, the two different calculations both specify the same (practical) resistor value. The difference between the simplified formula calculation and the complete formula calculation will be even less in the case of a germanium transistor.

Simple *current biasing* by the method just described has definite limitations, particularly in the case of germanium transistors. The more usual method adopted is voltage biasing with emitter feedback, this circuit being shown in Fig. 44. Here there are two effective bias resistors

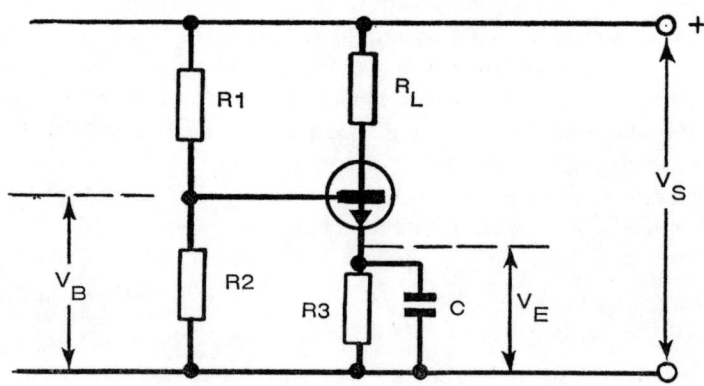

Fig. 44   Voltage biasing with emitter feedback.

UNDERSTANDING TRANSISTOR CHARACTERISTICS

R_1 and R_2, which work as a potential divider. There is also a resistor (R_3) in the emitter lead.

In this case the bias voltage developed (V_B) is given by:

$$V_B = \frac{R_2 \cdot V_S}{R_1 + R_2}$$

This ignores the voltage developed across R_2 by the base current, but the effect of this is negligible for normal calculation. The voltage at the emitter (V_E) is given by

$$V_E = V_B - V_{BE}$$

$$= \left(\frac{R_2 \cdot V_S}{R_1 + R_2}\right) - V_{BE}$$

The emitter current (bias current) is given by

$$I_E = V_E/R_E$$

$$= \frac{R_2 \cdot V_S}{R_E(R_1 + R_2)} - \frac{V_{BE}}{R_E}$$

To accommodate spread in transistor characteristics V_E should be large compared with changes in V_{BE}. Also R_E must be large to 'stabilize' the emitter current against variations in the supply voltage (V_S). In practice a voltage drop (V_E) of about 1 volt should be allowed in the case of germanium transistors, and a voltage drop up to 3 volts allowed in the case of silicon transistors, to stabilize the emitter circuit effectively.

On the other hand the values of R_1 and R_2 should not be so high that the base voltage is changed to a large extent by variations in base current. At the same time the values of R_1 and R_2 must be high enough not to waste power or drain power from the input signals. Logically the combined parallel resistance of R_2 and R_E should be less than the value of R_1. However, this would result in a considerable loss of input signal. This can be avoided by making R_2 about four times the impedance of the transistor.

A great advantage of this type of circuit is that provided the above conditions are met the operating current I_E is independent of *hfe*. Thus it can adjust automatically to any spread of transistor characteristics and maintain the desired collector current.

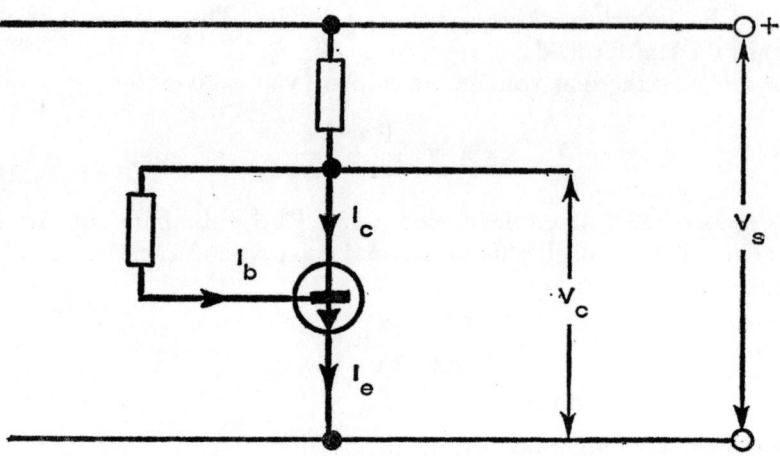

Fig. 45 Simple current biasing with collector feedback.

A third type of bias circuit which is sometimes used with silicon transistors is shown in Fig. 45. This is *current biasing with collector feedback*. In this case the collector voltage (V_c) is given by

$$V_c = V_s - R_1 \times I_E$$
$$= V_s - R_1 \times I_B (1 + hfe)$$

The base current (I_B) is given by

$$I_B = \frac{V_c - V_{BE}}{R_2}$$
$$= \frac{V_s - V_{BE}}{R_2 + R_1 (1 + hfe)}$$

The operating current of the transistor is then given by

$$I_{BE} = \frac{(V_s - V_{BE})(1 + hfe)}{R_2 + R_1 (1 + hfe)}$$

It should be noted that all the circuits described are for a *common emitter* configuration, which is the one normally used in radio circuits because of its superior characteristics as an amplifier. There are other types of bias circuits applicable to the common emitter configuration

UNDERSTANDING TRANSISTOR CHARACTERISTICS

and also two other configurations—*common-collector* and *common-base*. The *common-collector* configuration automatically provides compensated bias. With the *common-base* configuration, bias normally requires the use of two batteries.

Formula Check: Simple current biasing—Fig. 43

This is the simplest form of bias and it is only necessary to determine the value required for one component—R_1.

The important parameters are:

$$\text{base current } I_B = \frac{V_s - V_{BE}}{R_1}$$

It follows that the emitter current is:

$$I_E = \frac{V_s - V_{BE}}{R_1} (1 + hfe)$$

In this case the variations in working current (I_E) are substantially dependent on the spread of *hfe*, thus this method of bias is really only suited to transistors which have a low *hfe* spread.

The emitter current is also very much influenced by changes in supply voltage. However, this effect can be offset to a large extent by including a resistor R_2 of high value in the supply line, the higher the value used the less dependent the emitter current will be on the supply voltage.

Formula check: Current biasing with collector feedback —Fig. 45

With this form of bias the collector voltage is given by

$$V_C = V_s - I_B R_1 (1 + hfe)$$

The base current is governed by the difference between I_C and V_{BE} and resistor R_2, and can be determined as:

$$I_B = \frac{V_s - V_{BE}}{R_2 + R_1 (1 + hfe)}$$

The emitter current is given by

$$I_E = \frac{V_s - V_{BE}}{\frac{R_2}{(1 + hfe)} + R_1}$$

It follows that variations in the operating current (I_E) can be considerably influenced by spreads in *hfe* unless the value of R_1 is equal to or greater than $R_2/(1 + hfe)$, which in practice is normally only possible with a large supply voltage.

Formula check: Voltage biasing with emitter feedback —Fig. 44

With this form of bias the bias voltage (V_B) is determined by the values of resistors R_1, R_2 and the supply voltage (V_s), as long as the base current does not load the potential divider R_1, R_2.

$$V_B = \frac{R_2 \cdot V_s}{R_1 + R_2}$$

The emitter voltage is given by

$$V_E = \frac{R_2 \cdot V_s}{R_1 + R_2} - V_{BE}$$

The emitter current of the transistor is thus

$$I_E = \frac{R_2 \cdot V_s}{R_E (R_1 + R_2)} - \frac{V_{BE}}{R_E}$$

In order to nullify the effect of spreads of transistor characteristics, and changes due to temperature effects, V_{BE} must be kept as constant as possible, which means that V_E should be large in comparison with V_{BE}. Equally, to nullify changes in supply voltage affecting I_E, R_E must also be large. In practice this demands a value of V_E of about 1 volt for germanium transistors, and a V_E of about 3 volts with silicon transistors. This desirable mode of working is also achieved if

$$R_E \text{ is equal to or greater than } \frac{R_1 \cdot R_2}{(R_1 + R_1)(1 + hfe)}$$

This is the same as saying that for the emitter current to be independent of *hfe*

$$V_E \text{ should be equal to or greater than } \frac{R_1 \cdot R_2 \cdot I_B}{(R_1 + R_2)}$$

CHAPTER 7

AUDIO AMPLIFIERS

THE audio amplifier employed in the radio receiver comprises one or more stages of amplification coupled to an output stage for matching and powering a loudspeaker. Much depends on the *output power* required. An output power of about 0·005 to 0·01 watts (5 to 10 milliwatts) is high enough to operate high impedance headphones; and about 0·05 to 0·1 watts (50 to 100 milliwatts) a very small loudspeaker. Larger speakers to a portable transistor radio may require 0·2 to 1 watt output. Output power for conventional domestic radios is in the order of 2 to 5 watts; and for H-Fi, about 10 watts upwards.

The primary requirement of an amplifier is to produce power amplification or *gain* of the *af* signal. At the same time it should do this with minimum *frequency distortion* and maximum *linearity*. *Frequency distortion* occurs when the amplification is not the same on all frequencies, thus the frequency components are not represented at their correct relative strength (some frequencies being emphasized and others depressed). *Linearity* refers to the true reproduction of the actual waveform of the signals.

Frequency distortion is related to the gain of the amplifier. Ideally the relationship between gain and frequency should be a straight, parallel line. In practice this is impossible to achieve over the whole *af* range. In particular there will be a marked loss of gain at each end of a frequency range, i.e. at the lower frequencies and the higher frequencies —Fig. 46.

Provided the rest of the curve is reasonably linear, the addition of a *tone control* will normally satisfy most listening requirements. When more exact reproduction is required as in Hi-Fi, frequency distortion, where present, can be compensated by equalization circuits. These aim at introducing compensation of an opposite nature to smooth out the frequency response curve.

The type of distortion produced by *non-linearity* is usually more noticeable, and therefore less acceptable. As a general rule, the greater the gain extracted from a single stage, the greater the degree of distortion. No practical amplifier has exactly linear characteristics, even at low gain, but non-linearity shows up increasingly with increasing gain. Basically, therefore, a large number of individual amplifier stages, all operating at low or moderate gain should give less overall linear distortion than one or two stages operating at high gain. In practice it is necessary to adopt a compromise solution based on arriving at an

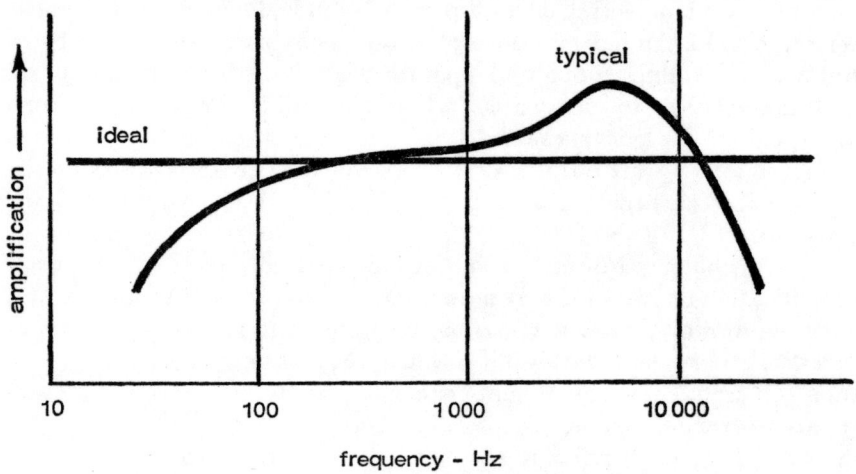

Fig. 46 Amplifiers do not amplify all frequencies by the same amount.

acceptable level of distortion in order to avoid an excessive number of separate stages. Additional stages can, in fact, introduce other troubles. In any case the question of distortion is not significant in the final output stage, where other forms of distortion can be present. Ultimately, too, the loudspeaker itself can be the main source of distortion in the whole system. It is a waste of effort to design an audio amplifier circuit with excellent linearity characteristics only to operate a poor quality loudspeaker from it. Conversely, a Hi-Fi quality loudspeaker cannot be expected to compensate for the distortion inherent with a poor quality audio amplifier.

For this reason it is really necessary to design the audio amplifier 'backwards', i.e. start with a selection of speakers and their requirements and work backwards to the first stage of amplification. This first demands some knowledge of the characteristics of amplifier circuits.

Class A Amplifiers

The simplest form of amplifier circuit is a single transistor with bias and input signal voltage such that the collector current always flows. This is known as Class A operation. It has the advantage of producing a low distortion as well as being simple to design and construct, but the disadvantage of drawing a relatively high current all the time.

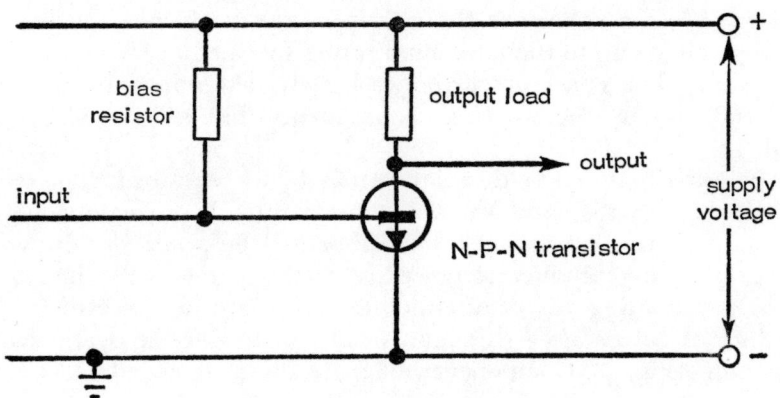

Fig. 47 Basic amplifier circuit using a silicon N-P-N transistor.

The simplest practical form of this type of circuit is shown in Fig. 47, using a single resistor to provide current biasing (*see* Chapter 6). The output load must be of relatively high impedance (several thousand ohms). This can be provided by a second transistor amplifier stage, or if used as an output stage by a step-down transformer to match the characteristically low impedance of a loudspeaker (4–16 ohms nominal) to the stage requirements. The same circuit can be used with a P-N-P transistor or an N-P-N transistor, the only requirement being that the supply voltage is opposite in polarity. Choice of an N-P-N transistor would be preferred as this type is more stable with simple current biasing using a single resistor.

It is more usual, and generally more desirable, to employ voltage biasing with emitter feedback, where the basic circuit is as shown in Fig. 48. Again the same type of load is used for the output, and the circuit is identical for a P-N-P transistor and an N-P-N transistor, with supply voltage polarity reversed. Values of suitable bias components can be determined from Chapter 6.

The *power-gain* which can be achieved from such a circuit depends on the transistor characteristics and the load. Transistor characteristics are given in the form of graphs of collector currents (I_C) plotted against collector voltage (V_C) for different values of bias (base current, or I_B). These curves typically take the form shown in Fig. 49 (*see also* Chapter 6).

The collector voltage is the supply voltage, which in theory can be any voltage up to the maximum rating specified from the transistor concerned. The collector current is largely determined by the load, and the bias by the working requirements (but *see also* Chapter 6 again).

The *power* that can be dissipated safely by a transistor is represented by the product of I_C and V_C, and is represented by a single value, i.e. so many watts. This can also be represented by a *load line* drawn on the graph. Any product of power *below* this curve is feasible to use (provided limiting values of either I_C or V_C are not exceeded). Any product of power *above* this line is not usable since it overloads the transistor—Fig. 50. Component values are therefore calculated accordingly.

Note: this also emphasizes that the maximum power rating of a transistor is not simply the product of I_C max and V_C max as it is sometimes supposedly taken to be. In other words it is *never permissible to apply both maximum V_C and maximum I_C to a transistor simultaneously as this will drastically overload it*, causing it to burn out.

A simple check on individual transistor characteristics will confirm this. For example for an OC84 transistor

$$V_C \text{ max} = 25 \text{ volts}$$
$$I_C \text{ max} = 500 \text{ mA}$$
$$\text{Power max} = 260 \text{ mW}$$

Erroneously employing V_C max with I_C max would give a power of $25 \times 500 = 12{,}500$ milliwatts—nearly 500 times the maximum power rating of the transistor!

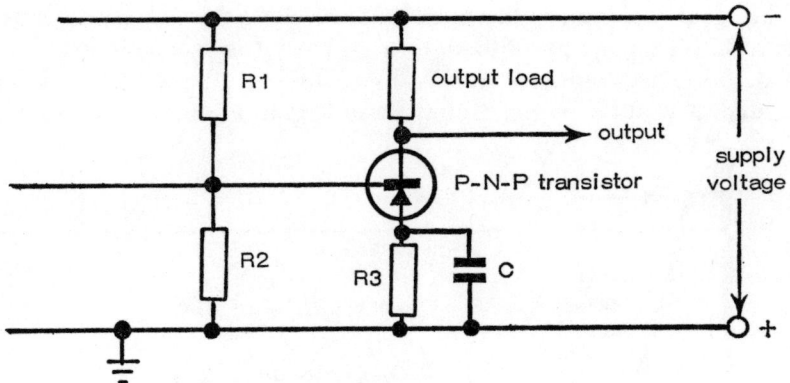

Fig. 48 Basic transistor amplifier with voltage bias provided by R1 and R2. R3 and C are stabilizing components. Polarity is shown for a P-N-P transistor.

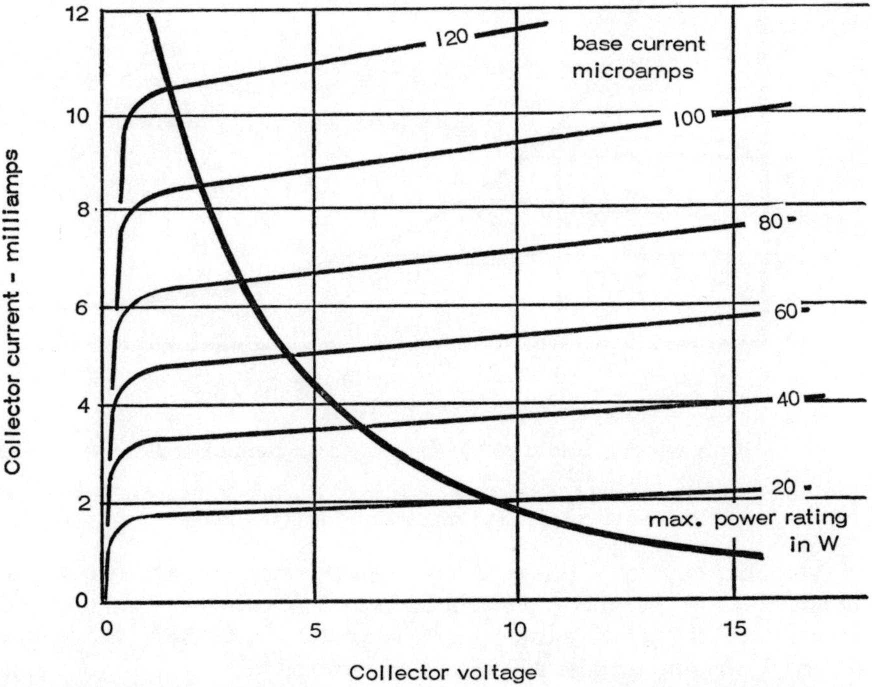

Fig. 49 Illustrating how maximum power rating of a transistor is related to collector current, collector voltage and base current in common emitter configuration.

The ideal maximum efficiency that can be achieved with Class A operation is 50 per cent, although in practice it is usually substantially less in order to avoid too much distortion, i.e. the amplifier is best operated at well below maximum possible gain to avoid troubles.

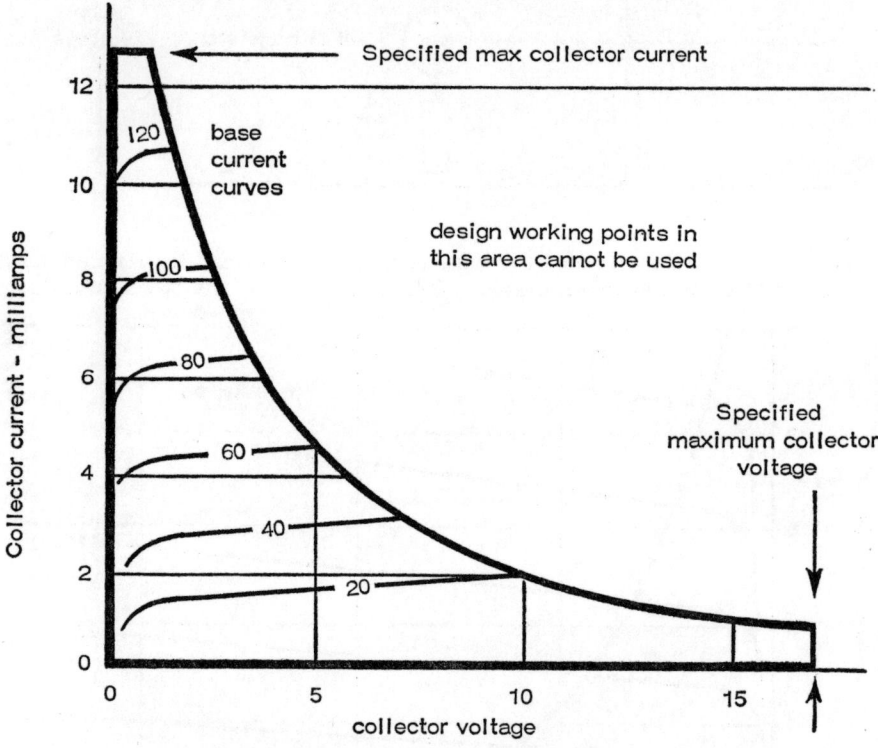

Fig. 50 Working envelope of a transistor as fixed by its power rating, maximum collector current and maximum collector voltage.

Amplifier circuits of this type can form the basis of amplifier stages connecting two or more circuits in series to derive the required amount of gain, whilst operating each stage under reasonably linear characteristics. Some simplification of circuitry is possible, rather than using complete voltage biasing circuits associated with each transistor. The output stage in a modern amplifier circuit, however, is normally a push-pull type operating as a class B amplifier. Push-pull circuits may also be used as a *driver* in multi-stage amplifiers.

Class B Amplifiers

The immediate advantage offered by a push-pull class B amplifier is that the output power obtained is considerably greater than double the power of a single transistor. Also the average current drain is very much lower than with Class A operation because the transistors are biased so that their working point is near cut-off and quiescent current is virtually zero. (Fig. 51 shows a 'working' diagram of Class B operating characteristics.)

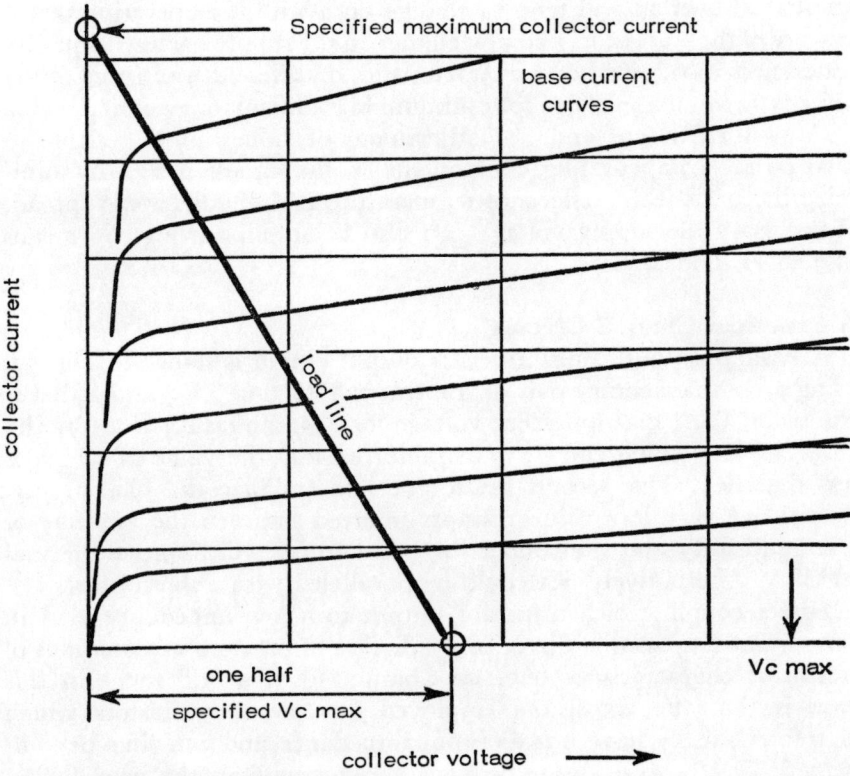

Fig. 51 Load line superimposed on transistor characteristics for a Class B amplifier.

Apart from much lower current consumption, making it more suitable for battery operated circuits, the efficiency of a Class B output can approach 80 per cent. It does have its inherent limitations, however, and in particular a proneness to *crossover distortion*. This is distortion produced at the change-over point when working is transferred from one transistor to another (changing from 'push' to 'pull' and 'pull' to 'push').

Crossover distortion will be most marked if both transistors are biased exactly to cut-off. It can be overcome, or at least the residual distortion can be substantially reduced, by selecting the bias so that, one transistor does not cut-off until the other has stopped conducting i.e. there is a slight overlap at the changeover. Unfortunately the amount of overlap will tend to change both with the operating temperature of the transistors and any change in the supply voltage. Equally, differences in the spread of characteristics of different transistors of the same type can make design for optimum bias difficult or even impossible without further 'cut and try' adjustment of values. It is possible to incorporate compensating components in the circuit design to minimize the undesirable effects of temperature and characteristic spread. If necessary the supply voltage can also be stabilized (e.g. by means of a zener diode).

A Practical Class B Circuit

A basic push-pull transformerless output circuit is shown in Fig. 52, using a complementary pair of transistors TR_2 and TR_3 and a driver transistor TR_1. Bias quiescent voltage for TR_2 and TR_3 is set by the value of R_1 (which can be a variable resistor) the value of which is less than R_2. This second resistor acts to stabilize the bias supply. Additional very low value resistors inserted between the emitters of TR_2 and TR_3 and the connecting point to C_2 will improve thermal stability. Alternatively, R_1 could be paralleled with a thermistor.

Direct coupling of a transistor output to a low impedance load in this circuit may seem a direct contradiction of previous explanations of transistor characteristics (e.g. *see* Chapter 6). The difference in this case is that the transistors employed are power transistors which characteristically have a *low* output impedance and can thus be connected directly to a low impedance speaker without the need for an output transformer.

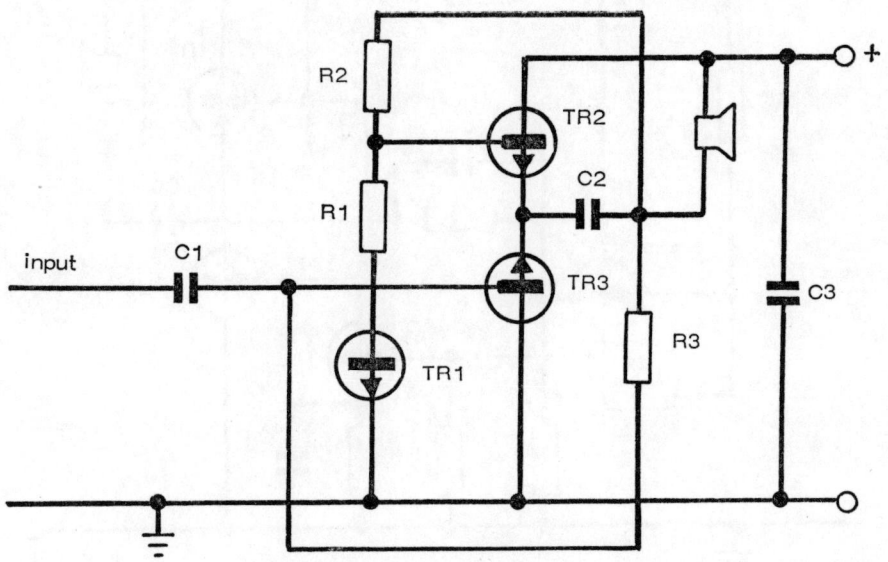

Fig. 52 Simple low power audio amplifier using a minimum of components with direct coupling to a miniature 4-ohm or 8-ohm loudspeaker. Component values:

R_1—68 ohm
R_2—1 k ohm
R_3—1 M ohm
C_1—8 µF
C_2—0·001 µF
C_3—50 µF
TR_1—BC 109
TR_2—AC 176
TR_3—AC 128
Supply: 9 volts

Conventional practice is to use this type of circuit with a single stage of initial amplification provided by a fourth transistor. This transistor can be made to work both as an amplifier and a *dc* difference amplifier, comparing the voltage offered by the potential divider with the 'end' voltage in the type of circuit shown in Fig. 53. Other points to note are that the negative feedback is taken from the output stage to the emitter of TR_1 via R_5, whilst the introduction of capacitor C_3 provides decoupling between these two points. A small amount of

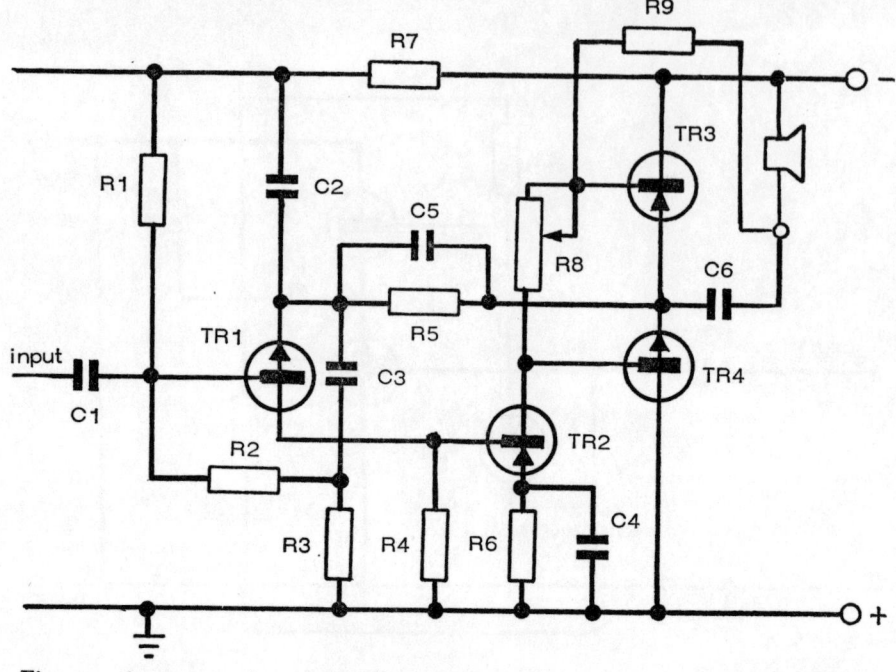

Fig. 53 A more or less standard type of circuit for a 1-watt audio amplifier.
Component values:
R1—15 k ohm
R2—15 k ohm
R3—2·7 ohm
R4—1·5 k ohm
R5—2·2 k ohm
R6—39 ohm
R7—2·2 k ohm
R8—1 k ohm
R9—510 ohm
C1—10 µF
C2—125 µF
C3—400 µF
C4—320 µF
C5—0·005 µF
C6—320 µF
TR1—AC 127
TR2—AC 128
TR3—AC 128
TR4—AC 127
Loudspeaker: 8 ohms
Supply: 9 volts

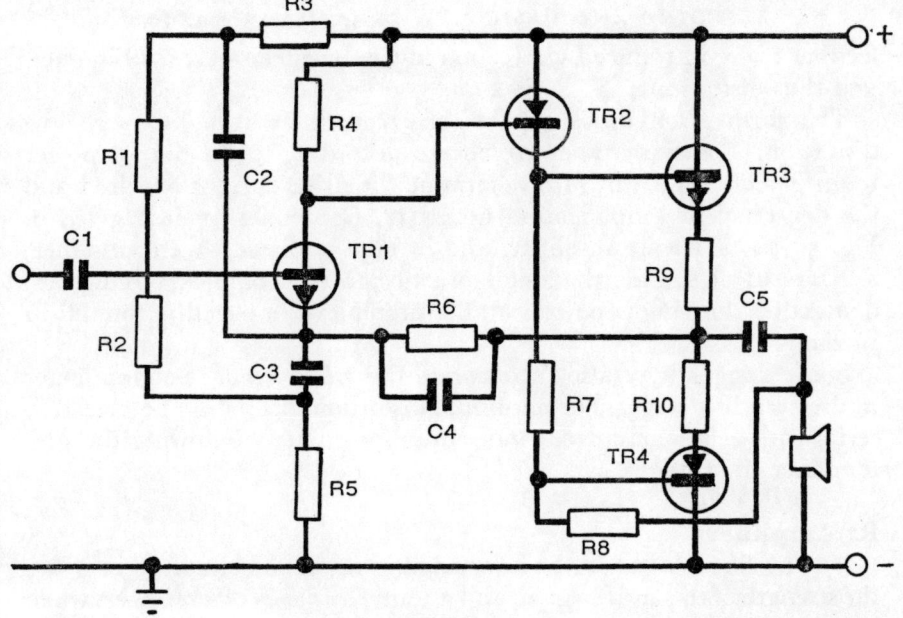

Fig. 54 A similar circuit to Fig. 53, but designed for 3 watts output, using higher power transistors. Component values:

R1—39 k ohm
R2—120 k ohm
R3—47 k ohm
R4—390 ohm
R5—22 ohm
R6—2·2 k ohm
R7—15 ohm
R8—560 ohm
R9—2·2 ohm
R10—2·2 ohm
C1—222 nF
C2—25 μF
C3—250 μF
C4—3·3 nF
C5—400 μF
TR1—BC 108
TR2—AC 128
TR3—AC 176
TR4—AC 128
Loudspeaker: 8–16 ohms
Supply: 22 volts

feedback is re-introduced via R3 to reduce the effects of spread in gain, and thus distortion.

This form of circuit is capable of excellent performance with low distortion. Transistor types are chosen according to the output power requirements, which in turn determine the drive current required and the degree of preamplification necessary. Component values given in Fig. 53, for a 1-watt amplifier, and in Fig. 54, for a 3-watt amplifier, are typical of 'standard' design practice. More complex circuits are demanded for high power audio amplifiers, especially for Hi-Fi working.

Some designs may also incorporate the *volume control* potentiometer in the amplifier stages. For minimum distortion this should be placed in series with a flow of current, normally immediately following the pre-amplifier (first stage).

Rf Amplifiers

An amplifier stage may be inserted after the tuned circuit to improve the strength of the aerial signal, and thus the *sensitivity* of a receiver, when it is known as a *preamplifier*. In this case, a basic amplifier circuit is adequate, e.g. similar to Figs. 47 or 48—but using an *rf* transistor. A somewhat superior performance, again with simple circuitry, is given by an FET amplifier—*see* Fig. 73, Chapter 11.

In practice, *rf* preamplifiers are mostly used in superhet receivers tuning to frequencies of about 50 kHz or above (e.g. FM radio circuits) because the signal-to-noise ratio inherent at these higher frequencies is usually poor without some amplification of the original frequency. An FET amplifier in this case is even more desirable than a conventional transistor amplifier because of its inherently lower noise level. Pre-amplifiers used in this context are generally known as *preselectors*.

CHAPTER 8

REGENERATIVE RECEIVERS

THE regenerative receiver works on a 'supercharging' basis, but quite different to the principle of operation of a superhet and very much simpler. Part of the *rf* reaching the *detector* is fed back and added as a 'boost' to the incoming signal, resulting in a substantial increase of signal strength. Effectively this increases the *sensitivity* of the receiver (as incoming signals are received at greater strength); and also increases the Q of the tuned circuit, and thus the *selectivity*.

This can be achieved by quite a simple circuit since the process of detection and regenerative feedback can be accomplished using a single *rf* transistor. In other words the transister is worked both as a detector (or rectifier) and an amplifier. There are just two basic requirements:

(i) The degree of amplification of *rf* achieved is proportional to the amount of feedback. It is limited only by the amount of feedback the transistor can handle without being excited into *oscillation*. Thus to achieve maximum gain it is necessary to introduce some sort of adjustment or control to give the maximum amount of feedback possible without the transistor being driven to the point of breaking into oscillation.

(ii) It is necessary that the feedback be *in phase* with the incoming *rf* signal, otherwise it will have a cancelling rather than amplifying effect on this signal.

The feedback signal can be coupled to the aerial circuit either *inductively* or by a *capacitor*. Inductive coupling is usually preferred, when the coupling coil can consist of a few turns of wire wound on a paper sleeve fitted to a standard tapped ferrite rod aerial coil—Fig. 55. The complete 'front end' of a simple regenerative receiver is then as shown in Fig. 56.

The choice of transistor is fairly non-critical, provided it is an *rf* type (not an *af* type designed for use in audio frequency circuits). Also

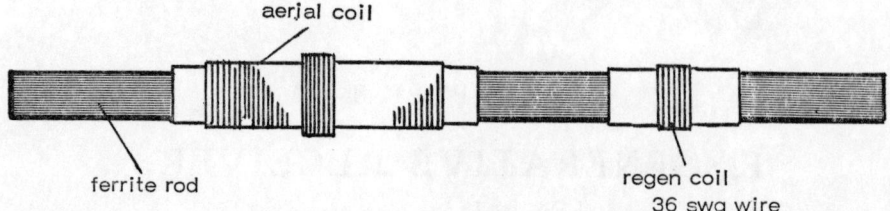

Fig. 55 Regen. coil takes the form of a few turns of enamelled wire on a separate sleeve free to slide on a ferrite rod aerial.

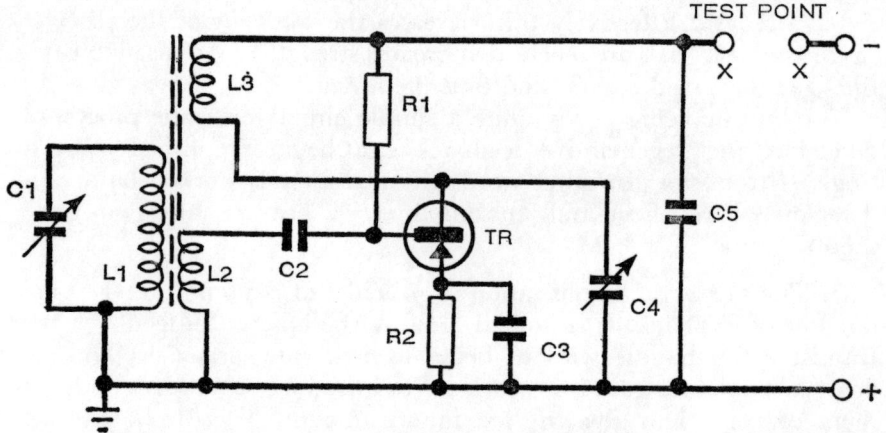

Fig. 56 Simple regenerative receiver circuit based on a P-N-P transistor. Test point marked for connecting high impedance phones to listen in and adjust circuit. For coupling to an audio amplifier, the output is taken from the collector of the transistor. Component values:

R1—1 M ohm (or higher value, as found best by test)
R2—3·3 k ohm
C1—tuning capacitor in standard aerial circuit
C2—220 pF (or larger, as found best by test)
C3—8 or 10 µF
C4—500 pF variable capacitor or trimmer
C5—8 or 10 µF
L1, L2—standard aerial coils on ferrite rod
L3—regen. coil (3–5 turns of 36 s.w.g. wire)
TR—OC45, or near equivalent
Supply: 9 volts.

the higher the performance characteristics the better. Current bias is provided by resistor R_1, the value of which is dependent on the choice of transistor type. Alternatively, using a 1 megohm potentiometer for R_1 will enable the bias to be adjusted to match the characteristics of a wide range of common low-power *rf* transistor types. L_2 is the coupling coil already referred to. R_2 and C_3 are the stabilizing components for the transistor, e.g. typically 1 to 3 k ohm for typical low-power transistors. C_2 can be any value between 2 μF and 10 μF (its actual value should not be critical). The remaining component is the 500 pF variable capacitor (C_4) included as a feedback control. Actually there are two feedback 'controls'—this variable capacitor and the coupling coil on the ferrite rod, the position of which can be altered.

The complete 'front end' can be coupled to a conventional *af* amplifier (e.g. *see* Chapter 7). However, it is not necessary to build a complete circuit to check that the front end is working. If the detector/regen. circuit is broken at points XX, and high impedance phones connected here, one or two strong stations should be received at listening strength, using a 9 volt battery supply. The original circuit shows a P-N-P transistor. Polarity must be reversed if an N-P-N transistor is used.

Checking out this circuit will establish that it is working properly and also shows how adjustments are made. Capacitor C_4 should be set to its mid position and an attempt made to tune into any broadcast station, adjusting only the tuning capacitor in the aerial circuit (C_1). Then adjust the *position* of the coupling coil on the ferrite rod to give maximum volume, readjusting the tuning capacitor as well, if necessary. If this is successful, C_4 should be adjusted forwards or backwards until the transistor starts to oscillate or 'howl', then backed off just enough for the 'howl' to disappear. Finally adjust C_1 again until maximum volume is heard. (Adjustment of C_4 will affect the tuning slightly, which is why C_1 has to be readjusted).

If no signal is received at all, then the feedback signal is probably out of phase. To correct this merely reverse the connections of the coupling coil. Failure to receive any signal could, of course, also be due to a poor aerial circuit used in an area where radio reception is weak.

If a signal is heard but no adjustment of C_4 produces oscillation, the coupling coil should be moved *nearer* the aerial coil on the ferrite rod to tighten the coupling. Alternatively if only a howl is heard with C_4 in its mid position, move the coupling coil *away* from the aerial coil until it disappears.

Once the optimum position for the coupling coil has been found it should be cemented to the ferrite rod. The optimum setting of the variable capacitor C_4 may need readjustment when an *af* amplifier is added to the 'front end' to complete a working radio receiver.

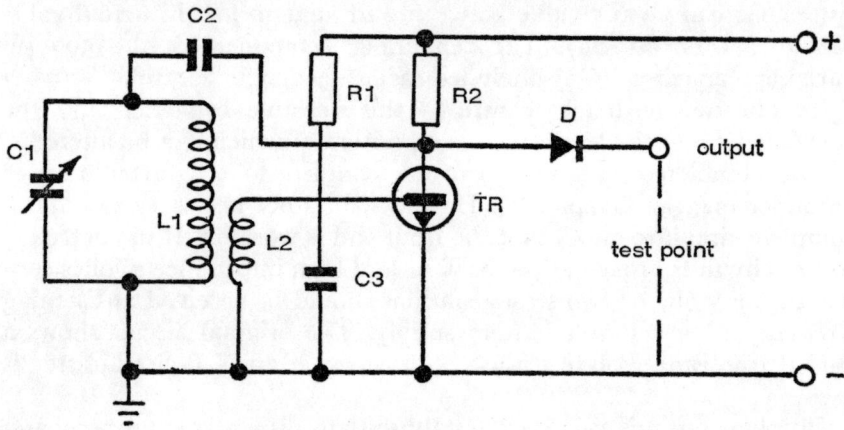

Fig. 57 Simple regenerative 'front end', not requiring a regen. coil. Test point marked is for listening in with high impedance phones to check circuit for working.

Component values:
R_1—1 M ohm
R_2—2·2 k ohm
C_1—tuning capacitor in standard aerial circuit
C_2—10 pF or 20 pF variable (trimmer capacitor)
C_3—0·01 µF
TR—2N2926 or equivalent N-P-N type
D—germanium point-contact diode
Supply: 9 volts.

An alternative design of regenerative 'front end' is shown in Fig. 57. This employs a transistor *pre-amplifier* followed by a conventional diode detector, a proportion of the pre-amplifier output also being tapped as feedback. In this case the feedback is capacitive—coupled to the aerial circuit via a small 10 pF or 20 pF 'trimmer' capacitor C_2. Again the transistor used must be an *rf* type.

Adjustment in this case is similar, but simpler. Again high impedance

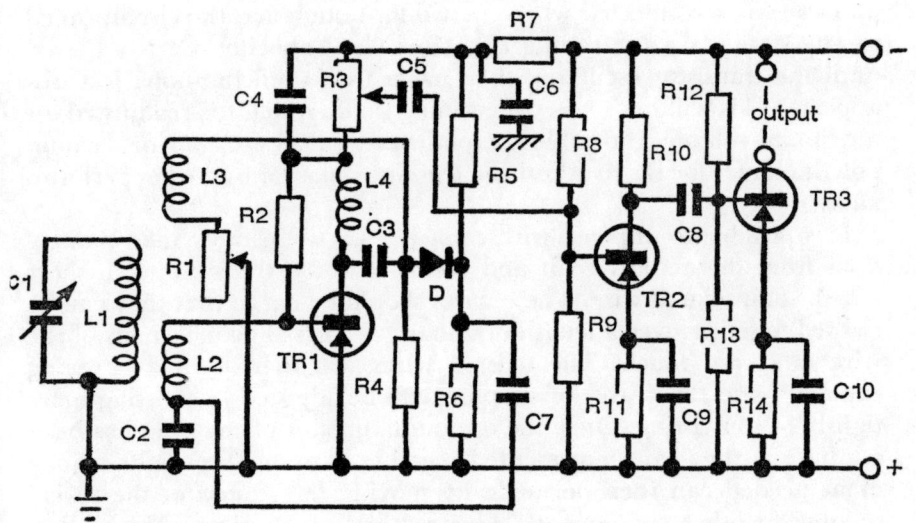

Fig. 58 Complete three-transistor regenerative receiver. A high impedance loudspeaker can be connected directly to the output. Alternatively the output points can be connected to a 10:1 output transformer, the secondary of which feeds a low impedance speaker. Component values:

R1—50 k ohm potentiometer
R2—560 k ohm
R3—5 k ohm potentiometer
R4—4·7 k ohm
R5—150 k ohm
R6—10 k ohm
R7—2·2 k ohm
R8—33 k ohm
R9—4·7 k ohm
R10—1.5 k ohm
R11—470 ohm
R12—680 ohm
R13—100 ohm
R14—10 ohm
C1—tuning capacitor in standard aerial circuit
C2—0·02 μF
C3—330 pF
C4—100 pF
C5—10 μF
C6—40 μF
C7—10 μF
C8—40 μF
C9—100 μF
C10—320 μF
D—germanium point-contact diode
L1, L2—standard aerial coils on ferrite rod
L3—regen. coupling coil on ferrite rod
L4—100 turns on 38-44 s.w.g. enamelled wire on $\frac{1}{4}''$ polystyrene coil former with iron dust core
TR1—AF 117
TR2—OC 71
TR3—OC 81
Supply: 9 volts

Note: regen. is adjusted by R1 and also by the core of L4. R3 works as a volume control.

phones can be connected where shown for testing and the circuit tuned to any station via the tuning capacitor C_1. Capacitor C_2 is adjusted until the transistor oscillates, then backed off until the howl just disappears. This will have upset the tuning slightly, so C_1 is readjusted for maximum volume. If trouble is experienced, substitute a 2 k or 5 k ohm potentiometer for R_1 to adjust the transistor bias for optimum performance.

It is possible to dispense with capacitor C_2 and simply take connections from the aerial circuit and collector of the transistor with short lengths of insulated wire. These wires should be cut so that they can be twisted together over a length of about $\frac{3}{8}$–$\frac{1}{2}$ inch. The bare ends of the wire must not touch. The twisted wires will then form a coupling capacitor and the degree of coupling can be adjusted by twisting more tightly or untwisting. Once the optimum amount of coupling has been established they can be cemented together. Any further slight adjustment needed can then be made by moving the *position* of the 'twist' relative to other components. This method of coupling, which eliminates one component, is quite critical of *position*, but despite its apparent crudity can work very well and is even used on commercial circuits.

A design for a complete regenerative receiver circuit using three transistors is shown in Fig. 58.

CHAPTER 9

THE SUPERHET

THE superheterodyne receiver or *superhet* works on the basis of combining the incoming *rf* signal with another high frequency signal and extracting a 'difference' signal at a fixed *intermediate frequency* (*if*). At the same time the *af* content of the original *rf* signal is transferred to the *if* signal frequency which now becomes a 'carrier'. The *if* output is then amplified at the intermediate frequency before being passed to the *detector*. Only the 'front end' of the receiver is effected—the detector and subsequent audio amplifier stages being the same as with simpler receivers except that the detector normally operates at a higher input because of the preceding amplifier stages.

This 'front end' comprises a *local oscillator* generating a high frequency equal to the *rf* carrier *plus* the intermediate frequency, and a *mixer*. The mixer receives *rf* from the aerial tuned circuit and *rf* plus *if* from the

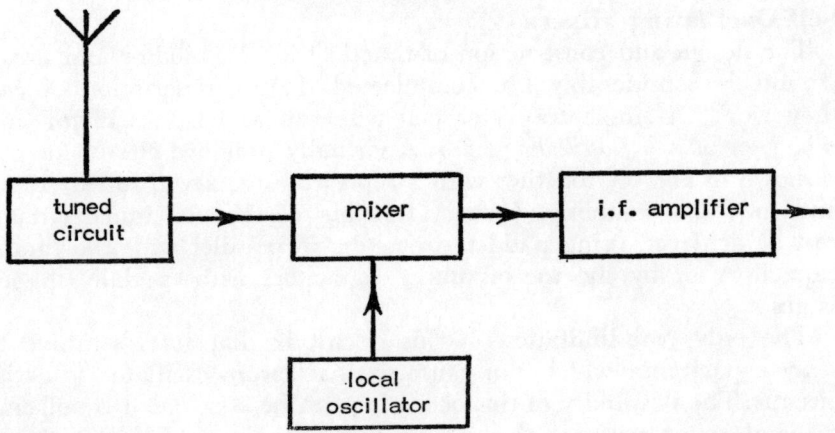

Fig. 59 The 'front end' of an AM superhet receiver in box diagram form. In the case of an FM superhet, the first 'box' is an *if* amplifier connecting to a dipole aerial.

local oscillator. From this it extracts the modulated *if* as an output, which is passed to the *if amplifier*—*see* Fig. 59. It follows that the local oscillator circuit must tune in step with the aerial tuned circuit, hence each incorporates a tuned circuit ganged together. The mixer receives a combined waveform as an input, the main components of which are the original aerial and oscillator frequencies together with their sum and difference frequencies. The *difference* frequency is then preferentially selected by the mixer tuned circuit and passed as an input to the *if* amplifier.

Exactly the same working principle applies for AM and FM receivers, except that the aerial signal for FM is normally received by a dipole aerial and *pre-amplified* before being passed to the mixer, (e.g. *see* Chapter 2). The other main difference is in the choice of intermediate frequency—470 kHz for AM and 10·7 MHz for FM. Also the use of more stages of *if* amplification is usually with FM (commonly three stages for FM as compared for two stages for AM). It can also be noted that with AM the local oscillator is (almost) invariably tuned to a frequency of 470 kHz *above* the aerial signal frequency. For FM, the local oscillator may be tuned to 10·7 MHz above *or* below the aerial signal.

Self Oscillating Mixers

The design and construction of practical local oscillator and mixer circuits is considerably less complicated than a description of *how* they work! A single transistor can act both as local oscillator and mixer, or as a *self-oscillating mixer*. A virtually standard circuit for this is shown in Fig. 60, together with a representative aerial tuned circuit and inductive coupling. Correct tracking of the two tuned circuits can be achieved using padder capacitors in parallel with the tuning capacitors, or by the use of tuning capacitors with specially shaped vanes.

The only real limitation of this circuit is that it is sensitive to stray capacitance which can cause feedback from oscillator to aerial circuits. The possibility of this occurring can be eliminated by placing a metal screen between the two sections, although this is not always a complete solution. Stray capacitance can also arise from leads, so all leads (particularly in the aerial section) should be kept as short as possible. If an external aerial is coupled to the aerial circuit, then it

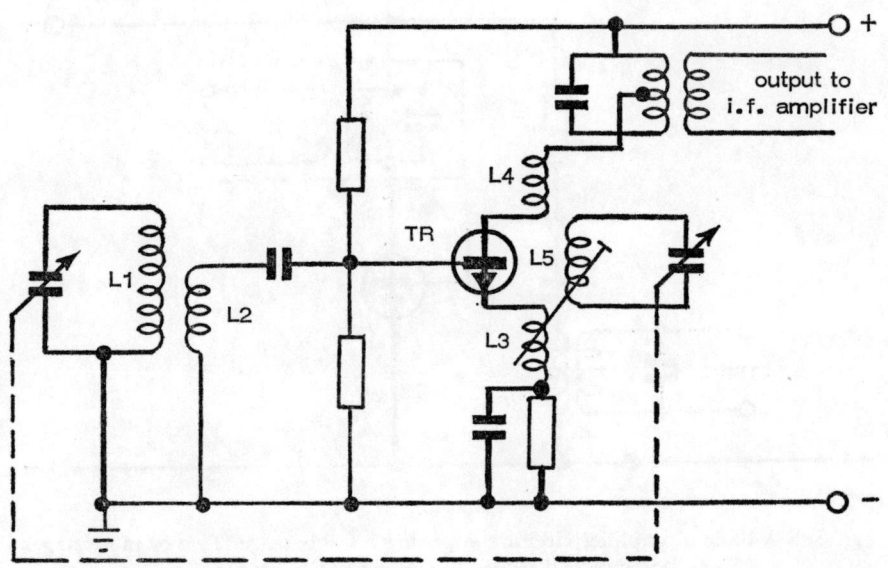

Fig. 60 This is a basic—and virtually standard—design for a combined oscillator-mixer circuit. Typical component values can be found in Fig. 65. L1 and L2 are the aerial coils. L3, L4, L5 are the windings of the oscillator transformer, L5 and its associated variable capacitor forming a tuned circuit tuned in step with the aerial tuned circuit by ganging the respective capacitors (as indicated by the dashed line).

should feed by a series capacitor to limit the extra capacitance that the aerial wire can add to the tuning capacitance in this circuit (*see also* Chapter 2).

If Amplifiers

An *if* amplifier stage usually consists of a transistor working as an amplifier with a tuned input and output circuit. The tuned circuits are provided by *if transformers*, tuned to the intermediate frequency. Thus in an AM/FM receiver, separate *if* transformers are required at each stage usually with primaries connected in series, and secondaries connected in series, ultimately feeding separate detector circuits.

A basic circuit for a single stage is shown in Fig. 61. The *if* transformer normally consists of a canned coil with an iron dust core for inductive

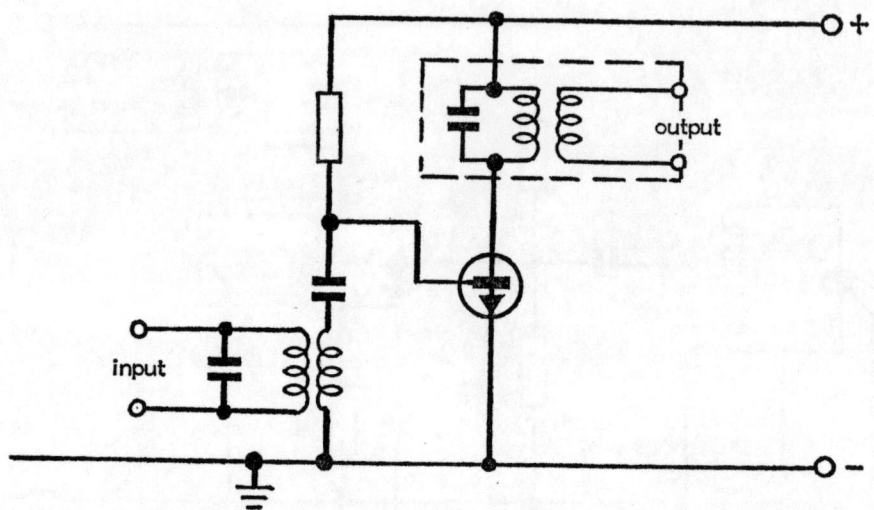

Fig. 61 A basic *if* amplifier circuit comprising a single stage. Two or more stages may be employed before the detector, e.g. *see* Fig. 65.

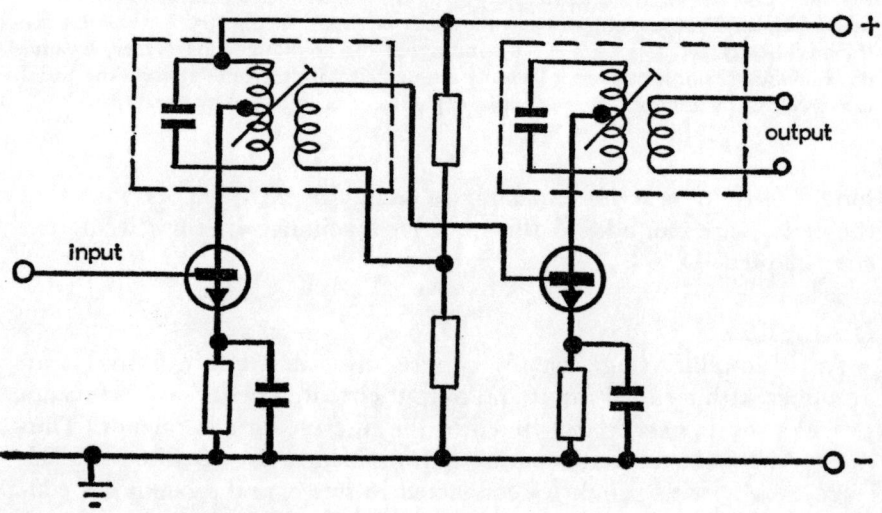

Fig. 62 Two stages of an *if* amplifier circuit, showing the standard method of coupling. The single-tuned *if* transformers (in dashed boxes) are adjusted by iron dust cores, to resonate at the *if* (intermediate frequency) chosen.

tuning. For best results, both coils of the transformer should be tunable.

Stage connections are straightforward and follow on the lines shown in Fig. 62. In high quality designs, however, filters may be interposed between stages. These may be piezoelectric crystals or ceramic filters manufactured to be resonant at the intermediate frequency. The use of such filters can improve *if* tuning and reduce noise. It should be noted that such filters should be of *bandpass* type, i.e. capable of passing the necessary *af* bandwidth and not single-signal crystals. A typical double-tuned two-stage *if* amplifier is shown in Fig. 63.

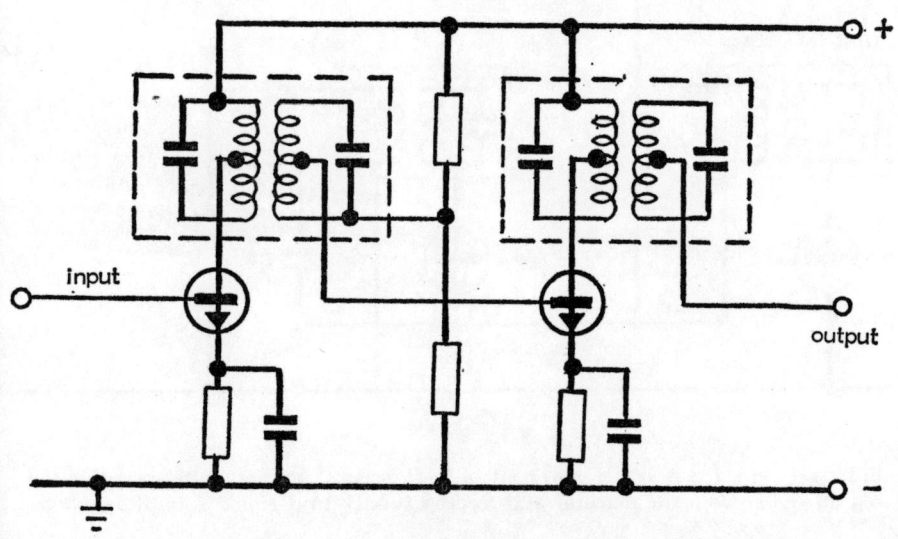

Fig. 63 Two stages of a typical *if* amplifier with double-tuned *if* transformers. Apart from both primary and secondary of the transformers being tuned, this circuit is virtually identical to Fig. 62.

Automatic Gain Control

The sensitivity of a superhet receiver is inherently high in the *rf* and *if* stages. This implies that in the presence of a strong AM signal excessive amplification can occur in the *if* amplifier, resulting in 'clipping' of the signal. Since *rf* signal strength received in the aerial circuit may vary enormously in strength from one station to another, the possibility of this happening is very real in practice.

Clipping, and thus distortion of the amplified signal, can be avoided by applying *automatic gain control* (*agc*) whereby a proportion of the *if* output is fed back to the front of the *if* amplifier to control the gain of the first stage. Any rise in signal *output* then automatically tends to reduce the gain at the input stage to compensate, and vice versa.

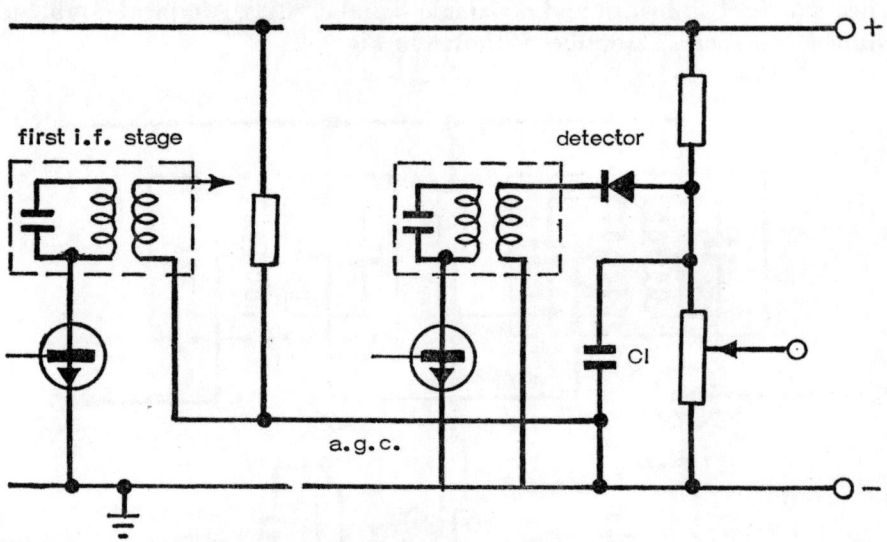

Fig. 64 Typical automatic gain control (*agc*) circuit 'tapping' a proportion of the output signal from the detector and feeding it back to the first *if* amplifier stage.

There are a number of ways this can be done, one of the simplest being shown in Fig. 64. A feedback signal is tapped from the detector and applied in series with the base-emitter voltage of the first *if* transistor, controlling the bias in proportion to the working voltage appearing at the detector. The change in bias thus modifies gain (i.e. reduces it with increasing detector voltage) in reverse proportion to the detector signal, i.e. provides automatic gain control. The resistor forms a potential divider feeding the base of the first transistor and can be chosen to bring the diode to its sharpest operating point, improving its response to weak signals. Capacitor C_1 is needed to eliminate *rf* and *af* voltages in the *agc* signal.

The presence of *af* signals in the *agc* voltage would reduce the modulation depth of the incoming signal. Equally, modulation must be filtered out so that the *agc* signal is in the form of a *dc* voltage with an average component following the strength of the carrier. This means that the time constant of the resistance and capacitance in the *agc* circuit must be long enough for this to occur, but not so long that the *agc* cannot follow rapid variations in detector signal strength, particularly fading. Normally this calls for a fairly large value of C_1, 8 μF being typical.

A diode can be added across the first *if* transformer although this is not strictly necessary. However it can substantially improve the gain control performance by acting as a damping device to make the circuit accommodate a wider range of *rf* input voltages.

Automatic gain control is not necessary with FM receivers since this function is already performed by the *limiter*. Sometimes it is added, however, to improve the performance of the *if* amplifier; or a single diode is added just ahead of the limiter to serve a similar purpose. What is virtually a 'standard' design of AM superhet circuit is shown in Fig. 65.

Superhet Alignment

A further problem always present with the construction of a superhet receiver is *alignment* of the *if* amplifier. This really demands the use of a calibrated signal generator as well as a universal meter. The signal generator should be capable of producing audio-frequency signals and also the intermediate and radio-frequencies it is necessary to explore, with provision for modulation at audio-frequencies. Where *af* and *rf* signal generators are employed the *af* generator should have an impedance of the order of 600 ohms, which can be fed directly into the top of the volume control of the receiver through a 10 K ohm resistor and a 1-μF isolating capacitor.

The *rf* generator should be of low impedance (of the order of 60 ohms output impedance) and, in general, can be fed into the base of the *if* transistor or mixer transistor via a 0·5 μF capacitor.

Where a signal generator is available, the *if* stages and signal circuit are aligned separately, in both cases using an output power meter or *ac* voltmeter connected across the speech coil (speaker terminals).

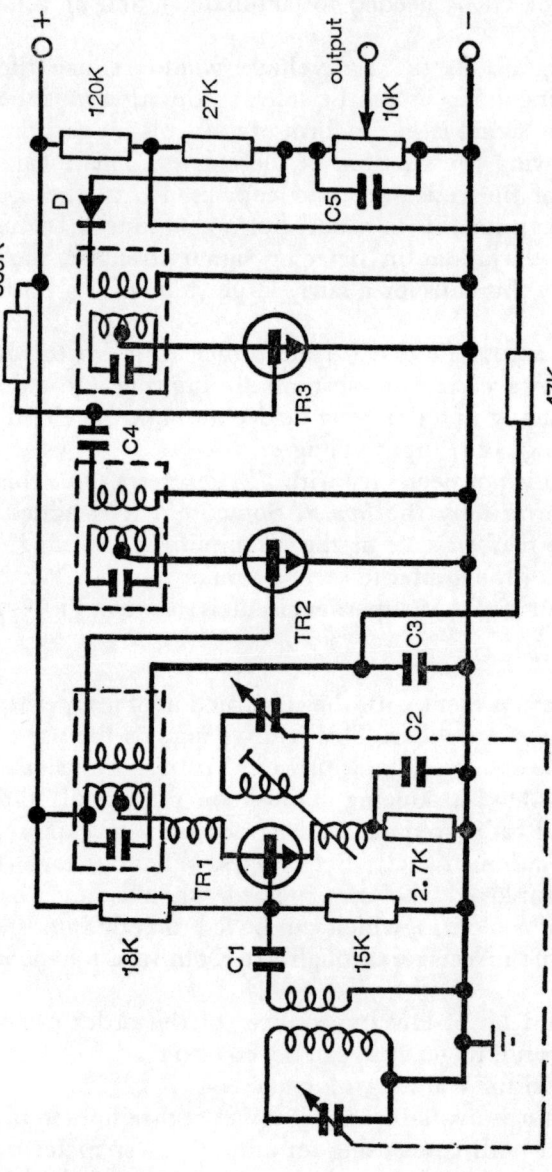

Fig. 65 This is virtually a standard design for the *rf* and *if* stages of an AM superhet receiver up to and including the detector. It also incorporates *agc*. Output is fed to an audio amplifier to complete the working circuit, the 10 k ohm potentiometer acting as a volume control. Resistor values are marked on the diagram.

Other component values:

C_1—10 nF
C_2—22 nF
C_3—1·6 nF
C_4—100 nF
C_5—10 nF
D—germanium point-contact diode
TR1—BF 194B, or equivalent
TR2—BF 195C, or equivalent
TR3—BF 195D, or equivalent.

The three *if* transformers used are single-tuned. Equally, double-tuned transformers could be used, modifying connections as in Fig. 63. The oscillator transformer is a standard type. Supply: 7·5 volts.

THE SUPERHET

Before attempting alignment the volume control should be set to a minimum in order to use the lowest signal from the signal generator consistent with a reasonable output, e.g. 50 milliwatts or 1 volt across the speech coil. This avoids *agc* action.

For aligning the *if* stages the signal is set at the *if* (normally 470 kHz) and usually applied to the base of the *if* transistors and mixer, in turn, working backwards (i.e. starting with the last *if* transistor). The corresponding *if* transformer cores are, in turn, adjusted to maximum output. The signal is injected using a 0·5 μF capacitor and an 820-ohm resistor in series with the generator output lead and never applied directly.

Having peaked the *if* transformers, the circuit is switched to medium wave, the generator set to a typical low frequency (e.g. 540 kHz) and the oscillator trimmer adjusted for maximum output.

The tuning capacitor is then set to minimum capacity and the second (mixer) gang trimmer adjusted for maximum capacity. These two stages are then retuned, as necessary, for optimum results.

For alignment of the signal circuits no direct connection is made, but the output or live terminal of the signal generator is connected to a loop of wire consisting of two or three turns approximately 7 to 9 inch in diameter with a series resistor in circuit of 430 or 390 ohms. The loop should be situated about 24 inch from the receiver ferrite rod. Normal procedure is then:

1. Set signal generator to a low m.w. frequency (e.g. 600 kHz) and tune gang condenser to the corresponding position on the scale. Adjust aerial coil position or aerial trimmer for maximum output.

2. Repeat with the signal generator set to a high m.w. frequency (e.g. 1400 kHz) and adjust gang condenser trimmer for maximum output.

3. Repeat operations 1 and 2 as necessary.

4. Switch to long wave, set generator to a typical frequency in about the middle of the l.w. band (e.g. 220 kHz) and adjust the longwave aerial coil position or trimmer for maximum output.

Where no signal generator is available, alignment must be carried out using broadcast stations as the source of signals. If an output meter with an impedance of around 25 ohms is available this should be connected in place of the loudspeaker; or alternatively a high-resistance voltmeter capable of reading 0–2 or 0–5 volts *ac* can be connected across the speech coil to give a visual indication of output signal strength. If neither type of meter is available, the output signal strength must be judged by ear.

1. Aerial and oscillator trimmers should be set to approximately the mid-point of their travel and the tuning control then set to the correct position for a local station, preferably in the middle of the m.w. band. Turn volume control to a maximum. Adjust the oscillator core to bring the station to the correct point on the tuning scale.

2. Turn the receiver so that it is oriented in the direction which gives *minimum* signal strength. Adjust *if* transformer cores for maximum output. (Note: if necessary, reduce the volume via the volume control to a minimum audible signal before adjustment as it is easier to judge an increase in strength of a weak signal audibly than an increase in strength of a strong signal. A meter will give a more positive indication regardless of the original signal strength, but at high signal levels the true output may be modified by the effect of *agc* action.)

3. Tune to the high-frequency end of the m.w. band, and adjust the oscillator trimmed to bring the station into tune consistent with the scale calibration. If this proves difficult or impossible, try adjusting first with the aerial trimmer followed by the oscillator trimmer.

4. Tune to the low-frequency end of the m.w. band. It should be possible to adjust for maximum signal by sliding the aerial coil along the ferrite rod. If not, it may be necessary to adjust the oscillator trimmer to get agreement with the scale calibration.

5. Steps 3 and 4 must then be repeated until an optimum adjustment is realized where adjustment of one end has no effect on the other.

6. Tuning on the long-wave band is not likely to be critical and it is usually only necessary to set the tuning condenser to the calibrated position and adjust the position of the long-wave aerial coil on the ferrite rod to bring in the station at maximum strength (or adjust the long-wave aerial trimmer, if fitted).

CHAPTER 10

INTERSTAGE CONNECTIONS (COUPLING)

A NY radio receiver consists of individual stages connected together, output to input, reading from the aerial to the loudspeaker (or conventionally left to right on a circuit diagram). The method of connection or *coupling* may be by direct connection (direct coupling); via a capacitor (*capacitive coupling*); or via a transformer (*inductive coupling*)—see Fig. 66 for diagrammatic representations. It is also necessary that

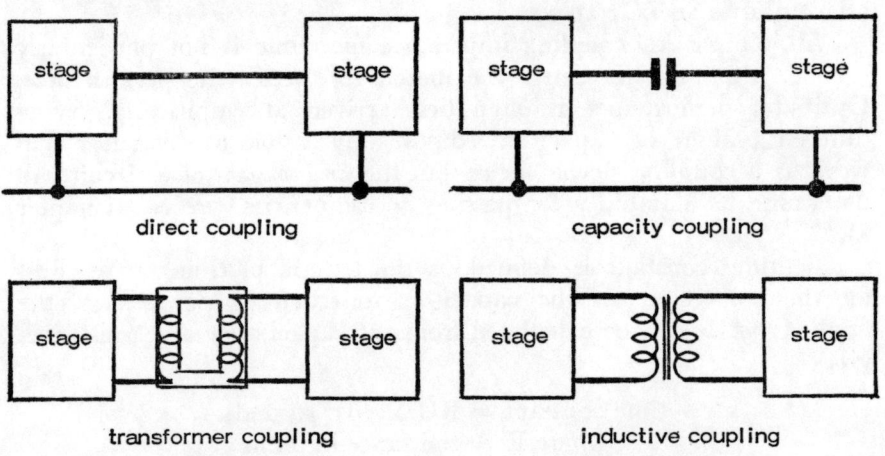

Fig. 66 Four different types of inter-stage coupling.

the output impedance of the 'feeding' circuit matches the input impedance on the 'receiving' circuit. This may dictate the type of coupling required. Alternatively impedance matching may be obtained by adjusting the load impedance of the 'receiving' circuit, varying the amount of coupling, or interposing an impedance-matching device.

The application of direct coupling is mainly limited to interstage connections where a varying *dc* voltage has to be passed on, e.g. the *af* output from a detector. It can also be used for interstage connections in an audio amplifier, or any coupling where 'matching' is correct and a *dc* as well as an *ac* path can be tolerated, or a *dc* path is necessary. Direct coupling has the obvious advantage of simplicity.

Both capacitance coupling and inductive coupling isolate output and input connections as far as the passage of *dc* is concerned. Capacitive coupling is accomplished with a series-connected capacitor. This is invariably associated with a load resistance and so is more correctly referred to as *RC* coupling. It has the advantage of using simple, inexpensive components and is widely used for coupling low-level amplifiers, etc., in radio circuits. It is not particularly suitable for coupling circuits carrying *rf*, when *inductive* coupling is preferred. In this the transformer (inductive component) is associated with a capacitor to 'tune' the coupling to the required frequency and is thus popularly referred to as an *LC coupling*.

With simple *RC* coupling impedance matching is not particularly critical, although the better the match the better the performance. Optimum performance is often best arrived at empirically, trying different values of capacitor. Almost any value of capacitor will work as a coupling device for *ac*, but the *time constant* of a circuit will determine its suitability for passing *ac* frequencies (*see also* Chapter 4).

The time constant is defined as the length, of time, in seconds, for the voltage across the capacitors to reach 63 per cent of the applied *emf*. It can be calculated from the capacitance and resistance, viz:

$$\text{time constant} = RC \times 10^{-6} \text{ seconds.}$$
$$\text{where } R = \text{resistance in ohms}$$
$$C = \text{capacitance in microfarads}$$

Specifically, the charge/discharge characteristics in terms of percentage voltage and time constant are presented in Fig. 67.

When it is necessary to accommodate a voltage applied at a particular frequency, one cycle of frequency implies a working of charge/discharge every *half* cycle. Thus taking a typical maximum *af* frequency of 10 kHz, the time to complete a full cycle is 1/10 000 or 0·01 milliseconds.

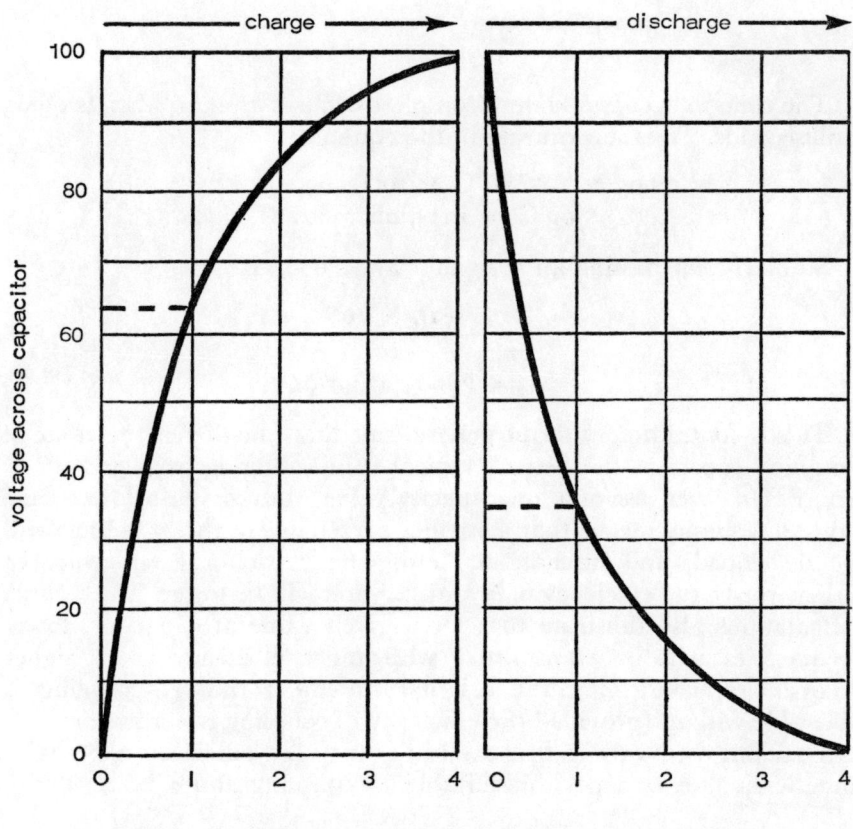

Fig. 67 Charge and discharge times or a capacitor expressed in terms of time constant. Thus approximately 100 per cent charge or 100 per cent discharge is achieved in 4 times the time constant.

Since *two* charge/discharge functions are involved, one positive and one negative, the true charge/discharge time is one half or 0·005 milliseconds.

Taking 1 000 ohms as typical of the input resistance for a low power transistor, i.e. base plus emitter voltage of 1 000 ohms (*see* Chapter 6), the time constant for RC coupling is:

$$\text{time constant} = 1\,000 \times C \times 10^{-6}$$
$$= C \times 10^{-3} \text{ seconds}$$

The time to accommodate a complete cycle of *af* at 10 kHz, is 0·005 milliseconds. Thus substituting in the equation

$$0.005 \times 10^{-3} = C \times 10^{-3}$$
$$\text{or } C = 0.005 \text{ microfarads or } \textit{less}.$$

Similarly, for passing an *rf* signal, say 10 000 kHz

$$C = \frac{0.15 \times 10^3}{10^6}$$
$$= 0.0005 \ \mu F \text{ or } \textit{less}.$$

This is contradictory to the general rule that 'the higher the value of coupling capacitor the better', typical values ranging from 500 nF to 10 μF. However, use of a low capacity value, thus giving a lower time constant, simply means that a smaller percentage of the *af* voltage will be developed, and discharged, across the capacitor each cycle. In other words, the efficiency of *ac* transmission will be lower. The sample calculations also illustrate that for a given value of capacitor, lower frequencies will be transmitted with more efficiency than higher frequencies, confirming the original statement that *rf* coupling is tolerable with *af* (provided the efficiency of coupling is not important), but seldom used with *rf* because of the very low efficiency likely with practical values of capacitors suitable for coupling duties.

Transformers

A transformer comprises, essentially, a primary winding and a secondary winding on a core of magnetic material. The actual form of the core governs the eddy current and hysteresis losses. The conventional transformer wound on a closed core, giving a continuous magnetic circuit, is not normally suitable for use with frequencies above about 20 kHz (i.e. above audio frequencies). Different forms of transformer are required for the inductive coupling of *rf* circuits if excessive power losses are to be avoided.

The *turns ratio* of a transformer—or ratio of number of primary turns to number of secondary turns—governs both the voltage transfer ratio and the current transfer ratio.

INTERSTAGE CONNECTIONS (COUPLING)

Secondary voltage $(V_s) = \dfrac{N_s}{N_p} \times V_p$

Secondary current $(I_s) = \dfrac{N_p}{N_s} \times I_p$

where N_p = number of turns on primary coil
N_s = number of turns on secondary coil
V_p = emf applied to primary
I_p = current flowing through primary coil
I_s = current flowing through secondary coil

The power transfer relationship is expressed by

$P_o = e \times P_i$

where P_o = power output from secondary
P_i = power input to primary
e = efficiency factor (which is always less than 1).

Normally a transformer is designed to have a maximum efficiency at a particular power output rating. Efficiency can then be expected to be lower both at lower or higher output powers. However *losses* will generally decrease with lower power, and increase with higher power. Maximum power rating is determined by the maximum temperature the windings and/or insulation can withstand, all energy losses in the transformer being dissipated in the form of heat.

The designation primary and secondary is one of convenience only. A transformer can be connected to work 'the other way round', if necessary, when the specified primary becomes the secondary and vice versa. Further types of transformers can also have a third or tertiary coil, e.g. *see* Chapter 4.

Impedance

The impedance relationship with a transformer is calculated on the basis of an ideal transformer without losses. This is generally satisfactory in practice provided the design of the transformer generates enough inductance to work with a low magnetic current at the voltage applied to the primary.

The basic impedance relationship then is

$$Zp = \left(\frac{Np}{Ns}\right)^2 \times Zs$$

when Zp = impedance looking into the primary connections
Zs = impedance of load connected to secondary

For *impedance matching* purposes, this formula can be rewritten

$$\frac{Np}{Ns} = \sqrt{\frac{Zp}{Zs}}$$

In this case Zs is the same as above (i.e. impedance of load connected to secondary); Zp becomes the primary impedance required to match; and Np/Ns the turns ratio required to produce this match.

A typical example of the application of this formula is the matching of an amplifier circuit requiring a high impedance output load to a low impedance load represented by a loudspeaker, using an output transformer. Suppose the load impedance required at the amplifier output is 5 000 ohms. This represents the primary impedance required. A typical loudspeaker impedance is 8 ohms thus, for matching

$$\frac{Np}{Ns} = \sqrt{\frac{5000}{8}}$$
$$= \sqrt{625}$$
$$= 25$$

In other words, the primary winding of the transformer should have 25 times as many turns as the secondary winding.

Inductive Coupling

Coupling by means of the mutual inductance between two coils or *inductive coupling* works on a similar principle to that of a transformer, but since any part of the magnetic field set up by one coil cuts the second coil the simple transformer formulas no longer apply. Also the degree of coupling achieved depends on the proximity of the two coils—the closer they are together the tighter the coupling, and vice versa.

The tighter the coupling the more the set-up operates as if the secondary were simply tapped across the corresponding number of turns on the primary—Fig. 68. If the input circuit is tuned to resonance, then the equivalent circuit is shown on the right where the resistor represents load resistance on the secondary coil with the coupling coils effectively working as an autotransformer.

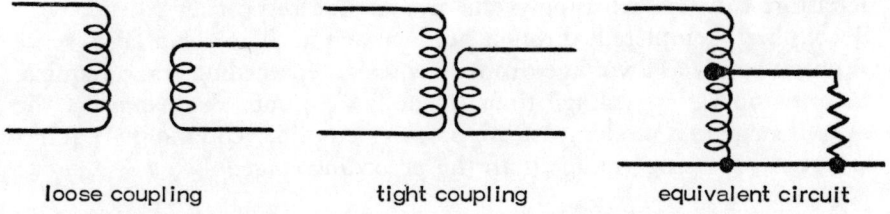

loose coupling   tight coupling   equivalent circuit

Fig. 68   The 'tighter' inductive coupling is, the more closely it approaches direct coupling in equivalent circuit (but providing a blocking path for *dc*)

In practice, by adjusting the number of turns on the secondary coil and/or adjustment of the tightness of the coupling, the impedance of the tuned circuit can be adjusted for optimum operation of the circuit for which the secondary is providing the input. For any degree of coupling, maximum transfer of energy is obtained when the reactance of the secondary coil is equal to the resistance of the load. Reactance can be calculated directly from the inductance of the coil and the frequency of the signal current, viz

$$\text{Reactance (ohms)} = 2\pi f L$$
where f is the frequency in Hz
and L is the inductance in henrys.

Decoupling

The object of *decoupling* is to prevent unwanted signals present in one circuit or stage of a receiver being introduced in another stage. Usually it is a preceding stage which may be effected by a *feedback*. Single stages are normally commonly connected throughout to provide supply and return lines from a single power source (i.e. the top and bottom lines in a circuit diagram). It is obvious that there is plenty of scope for the transmission of 'feedback'.

Separation or decoupling of one stage from the previous stage can be accomplished using a component offering high impedance to the unwanted feedback. It must be a conductive device to pass *dc* on to the preceding stage, but resistant *ac* to components at the frequencies that have to be eliminated. Thus it could be an RFC choke, or more normally, a resistor. This series connected component is then usually associated with a capacitor effectively paralleled across the supply—Fig. 69. Resistors can be surprisingly effective in this respect as 'filters', with decoupling completed through the capacitor. They can also serve simultaneously as a voltage-dropper when the preceding stage requires a lower operating voltage than the following one. For example, the output stage in a moderate or high power amplifier commonly requires a higher operating voltage than the preceding stage.

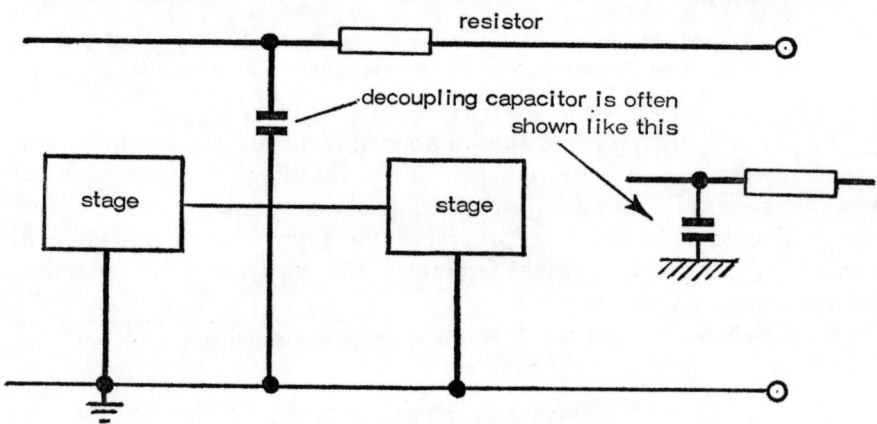

Fig. 69   Standard method of decoupling a preceding stage from feedback of unwanted signals.

In circuit designs decoupling components are often easy to identify by the fact that the capacitor is shown with an earth symbol immediately below it. This is done to avoid drawing a connection down to the bottom line of the diagram crossing a number of other lines on the way. The adjacent series connected resistor in the 'top' line will then be identified as the closely associated resistor (as in Fig. 69, right).

Similar considerations apply if the circuit design is drawn the other way up to accommodate reversed supply polarities. Conventionally the 'top' line in a circuit diagram is made the + line as regards the supply polarity, and the 'bottom' line the common earth or negative line. This convention cannot always conveniently be followed.

CHAPTER 11

FIELD EFFECT TRANSISTORS

FIELD effect transistors (FETs) are another group of semi-conductor devices which differ substantially both in construction and characteristics from conventional transistors. In fact, about the only thing they have in common is that their power requirements are low. Their actual working characteristics are more like those of a thermionic valve than a transistor.

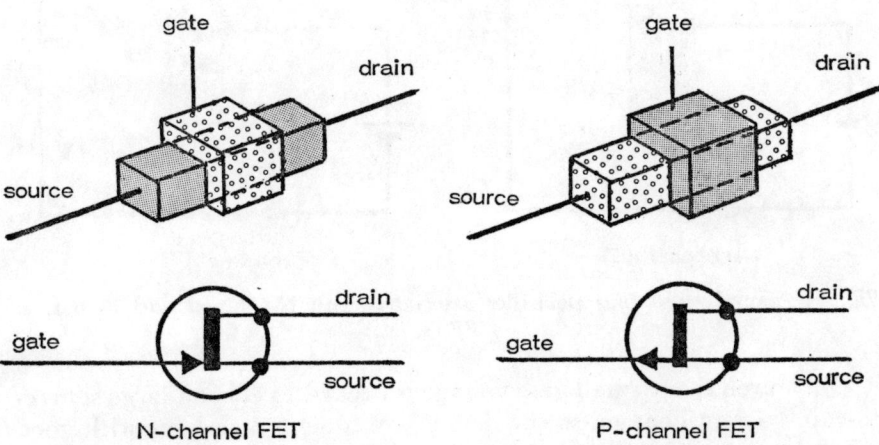

Fig. 70 Field effect transistors shown in diagrammatic and symbolic forms.

Construction, in effect, takes the form of a channel of N-type or P-type semi-conductor material enclosed for part of its length by a collar of opposite type material. One end of the channel is called the *source* and the other end the *drain*. The collar forms a *gate*—as shown in Fig. 70. This diagram also annotates the respective symbols for *N-channel* and *P-channel* FETs.

In simple terms, the FET works as follows. Application of a voltage across each end of the channel will cause a current to flow from source to drain (with electrons flowing from drain to source). Application of a voltage to the gate so that the gate is negative with repect to the source will reverse-bias the gate with respect to source, channel and drain, causing a depletion layer to be formed (*see* Chapter 5). This is equivalent to creating a constriction for the current flow from source to drain. The higher the gate voltage the greater the constriction, until eventually the source-drain current is reduced to almost zero. Connections of these two sources of voltage are shown diagrammatically in Fig. 71.

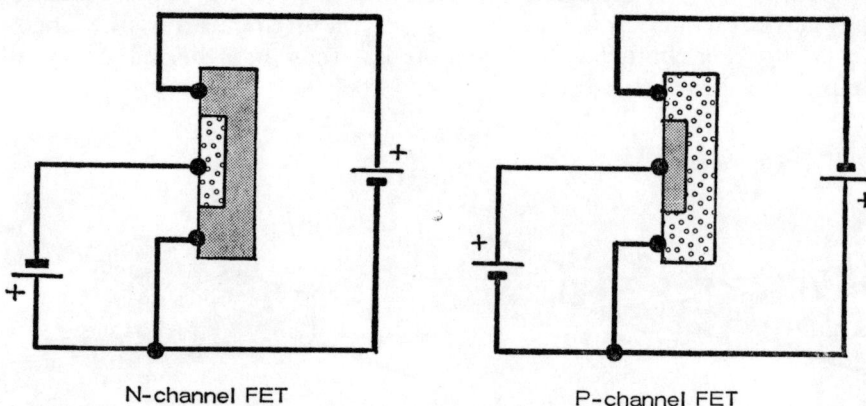

N-channel FET   P-channel FET

Fig. 71 Supply and bias polarities associated with N-channel and P-channel FETs.

Only a relatively small gate voltage is needed to effect a large source-to-gain current change, so the device works as an amplifier with good gain characteristics. Since the gate terminal is always reverse-biased, the resistance between the gate and the rest of the device is very high, i.e. an FET inherently has a high *input resistance*, usually the order of 1 to 10^3 megohms. In this respect it is similar in characteristics to a triode valve with the gate equivalent to the grid, the source equivalent to the cathode, and the drain equivalent to the anode.

It is also a characteristic of an FET that it has a very good performance with high frequency signals, making it particularly suitable for use in *rf* circuits. A further advantage for such circuits is that it has excellent low noise characteristics.

The type of FET described is, in fact, more correctly called a junction FET or JFET. There is another distinct type known as an insulated-gate FET or IGFET. As the name implies, the gate 'collar' is insulated from the channel by a very thin layer of insulating material. This has an even higher input impedance, typically of the order of 10^6 megohms or more. It is also a more flexible device since either a positive or negative voltage can be applied to the gate. With a JFET, the polarity of the drain supply is negative for a P-channel device and positive for an N-channel device, with the gate always reverse-biased (i.e. opposite polarity to that of the drain).

Field effect transistors are also made with two gates—*a signal gate* and a *control gate*. These gates are effectively worked in series. Such devices find a particular application in *rf* and *if* amplifiers, and mixers. They may be called MOSFETs, although this description is also applied to single-gate IGFETs.

It can also be mentioned that FETs may be classified as *depletion mode* or *enhancement mode* devices. The difference really refers to the effect of applying the gate voltage. *Depletion mode* reduces or depletes the current flowing through the channel. *Enhancement mode* increases or enhances the current flowing in the channel. Thus JFETs always work in the depletion mode, whereas IGFETs can be made to work either in the depletion mode or enhancement mode, depending on the polarity of the gate voltage. Either types of FETs can be constructed which have *no* channel as such, and thus no current flows with zero gate voltage. Application of a gate voltage then causes a channel to be formed, through which current flows. These are therefore *enhancement mode* devices.

Working Characteristics

Output characteristics of an FET are given in terms of drain voltage V_D and drain current I_D related to different values of gate voltage V_G. The first two descriptions are synonymous with channel voltage and channel current respectively, for which the alternative abbreviations are sometimes used.

V_{DS} = channel voltage (or voltage between drain and source)

I_{DS} = channel current (or current between drain and source)

Gate voltage (V_G) is also sometimes quoted as V_{GS} (gate-to-source voltage).

Characteristic curves are similar to those of conventional transistors—Fig. 72. The main difference is that the curves tend to flatten out rather less gradually, 'bottoming' voltage is higher, and drain voltages used tend to be higher than transistor collector voltages.

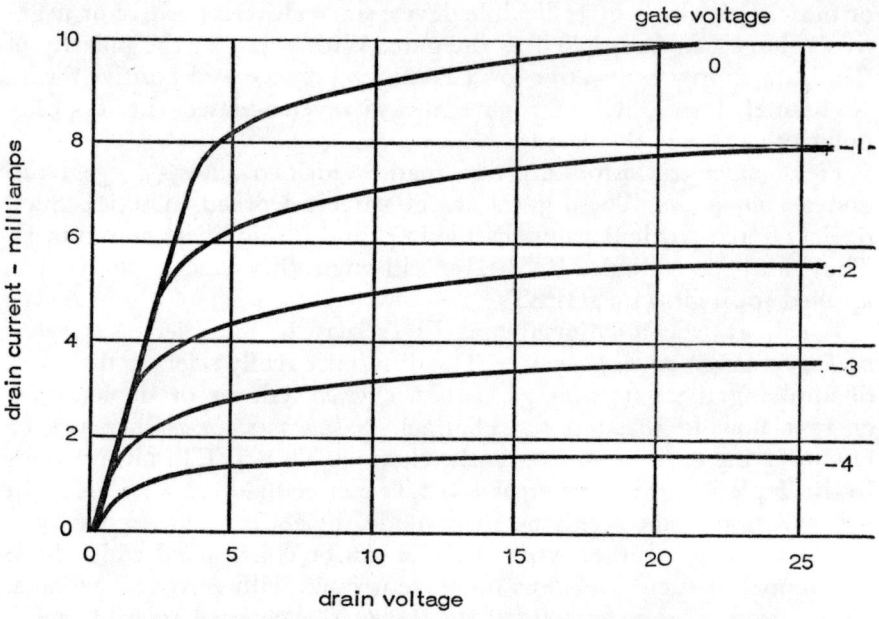

Fig. 72 Characteristic curves of a small FET.

Mutual conductance characteristics, as expressed in terms of channel current/gate voltage curves, are essentially similar to those of valve anode current/grid voltage curves.

FET Circuit Design

The following description applies to FETs working in the depletion mode, where the gate voltage is always of opposite polarity to that of the drain. A basic working circuit is shown in Fig. 73, where RG is the gate bias resistor; RS is the stabilizing resistor associated with a capacitor C2 to form a stabilizing load; and R1 is the drain resistor or drain load.

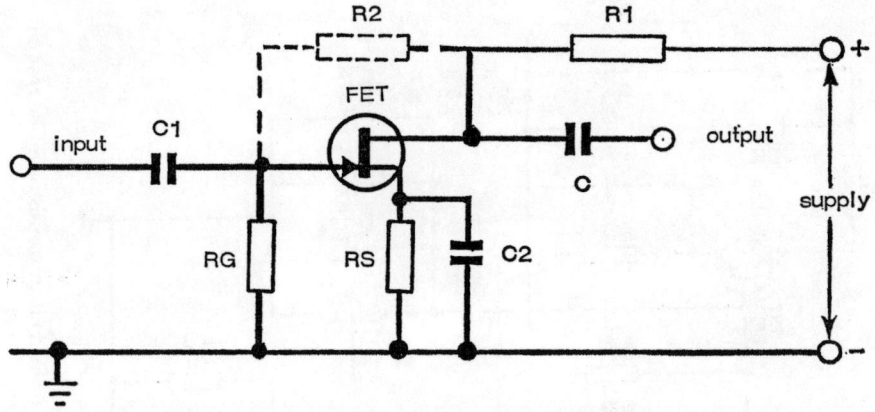

Fig. 73 N-channel FET amplifier circuit. With a P-channel FET the supply polarity would be reversed, but the circuit otherwise identical. Required component values are discussed in the text.

The required (negative) gate bias can be set by selecting the value of RS to bias the source positively to the required extent. This will also tend to stabilize the channel current against any spread of FET characteristics. Further stabilizing effect is produced by capacitor C2 in parallel. Effectively, the higher the value of RS the better in providing stability, although this will be at the expense of a loss of voltage through the channel. Also, and even more significant, is the fact that excessive gate bias may be developed in this way. Hence RS is generally restricted to a moderate value, although R2 could be added as shown to form a potential divider to limit the gate voltage to the required value. This is not a particularly attractive solution as the value of R2 needs to be substantially greater than RG, which itself is normally of the order of 10 M ohm.

With negative bias, gate current is normally zero. The purpose of RG is to hold the gate at normal zero potential under these conditions, hence the high value. Equally, capacitor C1 is necessary to prevent any direct *dc* voltage reaching the gate from the input. If the signal source contains no such direct voltage, then neither C1 or RG are necessary. An example of this is when an FET amplifier stage follows a tuned aerial circuit, when it can be directly coupled to that circuit if preferred.

An example of a high-performance circuit utilizing the favourable

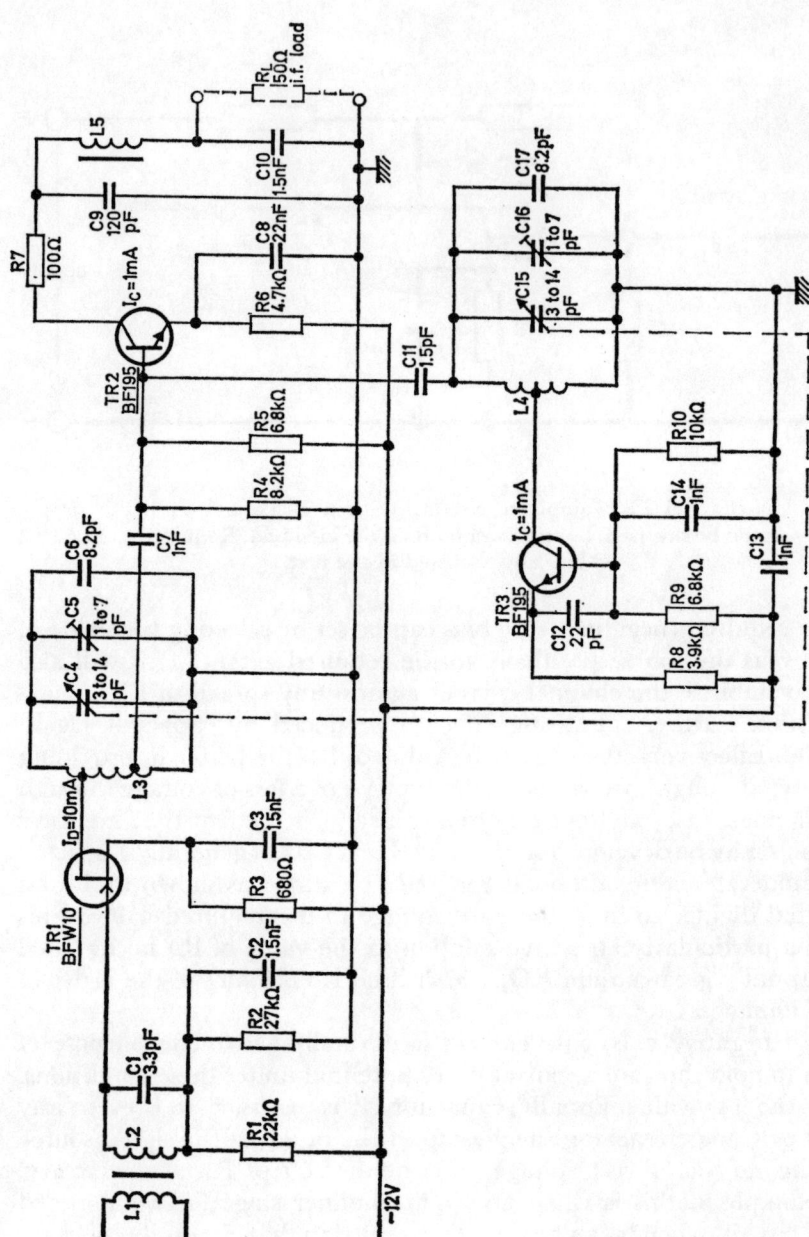

Fig. 74 Mullard design for a capacitively-tuned FM tuner using a BFW10 field effect transistor in the rf input stage. All component values are shown on the drawing.

characteristics of an FET is shown in Fig. 74. This is a Hi-Fi FM tuner circuit (by Mullard), and also illustrates application of the basic FET circuit design features just described. The aerial input circuit is transformer-coupled to the input of the FET, dispensing with the need for a coupling capacitor. *Dc* stabilization is provided by a voltage divider for the gate and a source resistance of 680 ohms. The following circuitry is a conventional mixer-oscillator, which in turn would be followed by an *if* amplifier and audio amplifier.

CHAPTER 12

MISCELLANEOUS CIRCUITS

Tone Controls

In its simplest form a tone control is an elementary tuned circuit comprising a resistor and capacitor in series applied across the primary of the output transformer, or the output. It is designed to be 'tunable' over the audio frequency range to be covered, the resistor normally being the adjustable component.

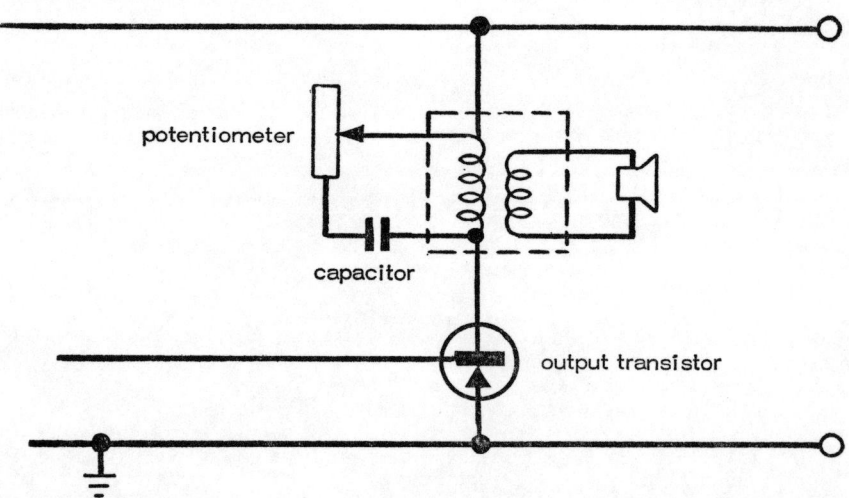

Fig. 75   Elementary tone control circuit applied across the output transformer.

Either a 5 k ohm or 10 k ohm potentiometer can be used for the resistor, associated with a capacitor of about 1 pF to 10 pF. This will give a reasonable tone control effect of audio frequency signals up to the order of about 10 kHz—Fig. 75. Its effect on the quality of reproduction will be fairly poor, but that is not necessarily significant with a small, low-power AM receiver with a small loudspeaker.

MISCELLANEOUS CIRCUITS

A better form of simple tone control is shown in Fig. 76, consisting of a series resonant circuit for treble cut and a parallel resonant circuit for bass cut. Each has its own separate control via a potentiometer. Mid-frequencies should be little affected by adjustment of either or both treble and bass controls.

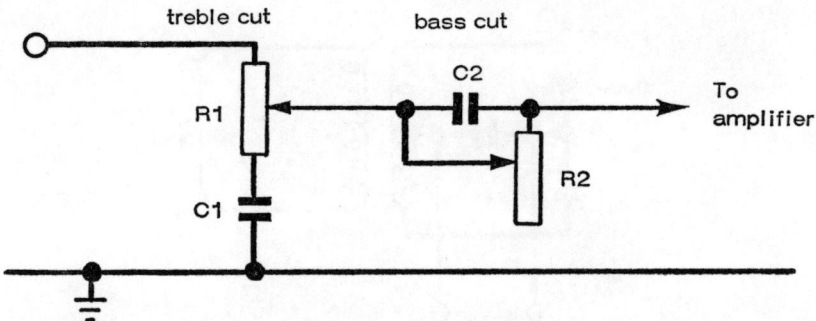

Fig. 76   Practical tone control circuit with separate treble and bass controls. This would normally be positioned in front of the audio amplifier.

The most suitable position in a receiver for a tone control of this type would be at the input of the audio amplifier (i.e. between a detector and amplifier), and it is especially suitable for use with conventional 4-transistor audio amplifier circuits commonly used on low and medium power receivers which employ feedback over two or more stages. It is a 'passive' type of control and thus cannot introduce instability in the following amplifier circuits.

Considerably more elaborate tone controls are used in Hi-Fi circuits, the object being to achieve a substantially flat response curve over the middle frequency range with increasingly progressive boost and cut towards each end frequency. The circuits may be of 'passive' type (e.g. Baxandall), or 'active' tye. The latter are normally based on frequency-dependent feedback networks between the base and collector of a transistor and therefore cannot be used in preceding circuits which themselves incorporate feedback without risk of introducing instability.

Power Supply

It is often more convenient—or economic—to power a transistor radio from the mains rather than dry batteries. This requires a separate *power supply* circuit (which can be built into the radio circuit if required), capable of supplying the required *dc* circuit voltage and maximum current needed. In theory such a duty can be provided by a step-down transformer with full wave rectification added, together with a smoothing capacitor—Fig. 77. A bridge circuit is normally used, however, as being less wasteful of input power.

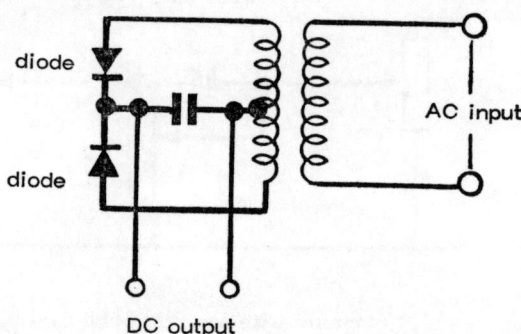

Fig. 77 Simple half-wave rectified *dc* supply from a step-down mains transformer.

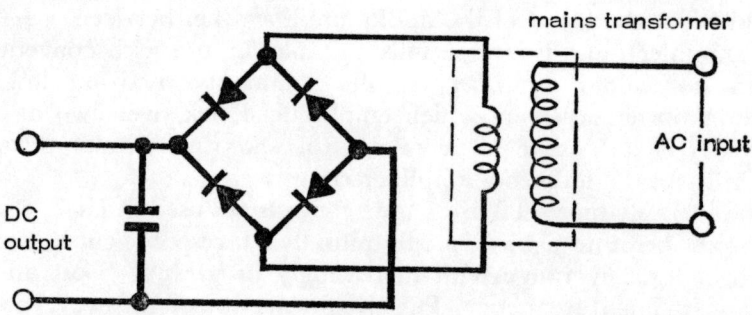

Fig. 78 A more practical form of *dc* power supply. Diodes are type BY164, or equivalent.

A typical power supply circuit is shown in Fig. 78, the diode rectifiers being chosen according to the actual power requirements, i.e. capable of carrying the maximum forward current in the output circuit. There will be some voltage multiplication in this circuit, typically of the order of 1·3×, so the transformer ratio is chosen accordingly. For

MISCELLANEOUS CIRCUITS

example, if a 12-volt output is required the 'voltage' multiplication characteristics imply that transformer output voltage $\times 1\cdot 3 = 12$, or a transformer step-down ratio should give $12/1\cdot 3 = 9\cdot 23$ volts, say 10 volts. In practice the actual output voltage is not critical, provided it or the resulting load current does not exceed the ratings of the semiconductor devices powered by this circuit. Neither is the choice of smoothing capacitor value particularly critical, provided it is of the order of 200 μF, or greater per watt output.

Tuning Meter

A *tuning meter* or signal meter indicates the strength of the signal into which it is being tuned. It also serves as a tuning *indicator*, maximum meter reading indicating the optimum tuning for that particular station.

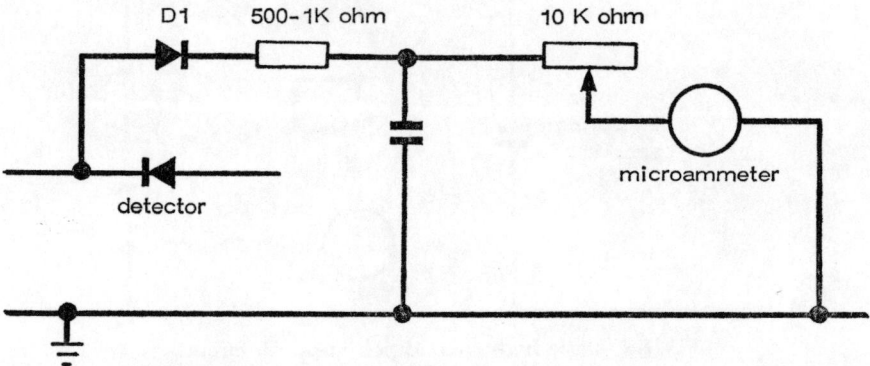

Fig. 79 Tuning indicator circuit. The diode (D1) can be any type (germanium point-contact preferred).

The obvious point to measure a signal strength is at a 'tapping' point in the detector stage, so arranged that it draws only a minimal current and thus does not degrade the signal strength. A suitable high resistance circuit is shown in Fig. 79, feeding a microammeter. A small proportion of detector current is tapped by this circuit, rectified by the diode D1 and further filtered by the resistor and capacitor. High circuit resistance is provided by the 10 k ohm potentiometer, which further acts as an initial adjustment to set the meter at zero with the set switched on, but in the absence of a signal. The meter itself should have a 0–10 microamp range. A miniature edge-type meter is usually preferred for compactness.

It is also possible to 'tap' the *agc* line for a tuning meter current (e.g. in superhets), only in this case it is usual to employ a transistor to amplify the variations in *agc* voltage and measure the changes in the collector current.

Battery Condition Indicator

Battery condition can be indicated by a paralleled circuit comprising a resistor and a milliammeter or microammeter—Fig. 80. The value of the resistance and meter range is chosen to give a full range reading at

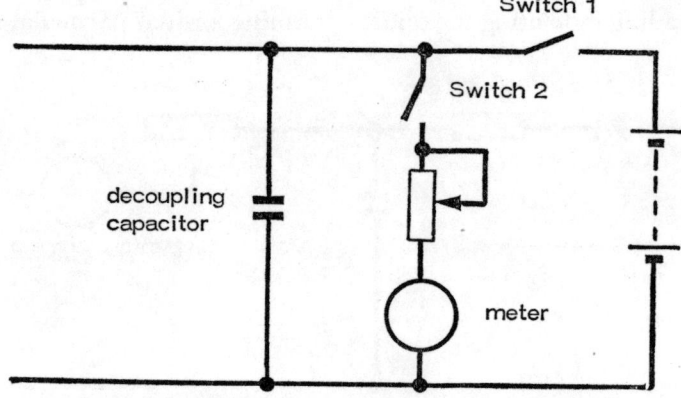

Fig. 80 Basic battery-condition indicator circuit.

maximum battery voltage, this reading falling as the condition of the battery deteriorates. The value of R can thus be calculated from:

$$R = \frac{\text{battery voltage}}{\text{maximum rage of meter in amps}}$$

Thus, using say a 0–1 milliammeter associated with a 9-volt battery

$$R = \frac{9}{0 \cdot 001}$$
$$= 9\ 000 \text{ ohms.}$$

This high resistance circuit, drawing only 1 milliamp at a maximum, could be left in circuit continously. It is generally advisable, however, to isolate the battery condition circuit with a separate switch S2. It is then only operative, i.e. indicating battery condition—when S2 is depressed. With this arrangement it is also practical to use a lower resistance circuit, e.g. to accommodate a 0-5 milliamp meter.

It should be noted that battery voltage under load (with S1 and S2 closed) will be less than the nominal battery voltage. Once the battery voltage falls to less than 80 per cent of the nominal voltage under load in the case of carbon-zinc cells, the battery has deteriorated to a stage where it should be replaced. Lowering of supply voltage in a transistor receiver normally introduces distortion, further aggravated by turning up the volume control to compensate for the loss of output signal strength.

A more practical form of battery condition indicator is, therefore, to use a potentiometer for R (e.g. a 10 k ohm potentiometer in the example above), so that with a fresh battery the meter can be adjusted initially to give full scale reading under load. The meter scale is then marked at 80 per cent full scale—at 0·8 milliamps in the case of a 0–1 milliamp meter. As long as the meter reads in the range 80 per cent–100 per cent full scale reading (i.e. between 0·8–1·0 milliamps in the example quoted), the battery is in 'good' condition. Any reading below the 80 per cent mark indicates that the battery needs replacing.

Impedance Matching Networks

The simplest type of impedance matching is by an L network. The left hand circuit shown in Fig. 81 is applicable when the desired impedance R_{in} is greater than the load resistance R. The second

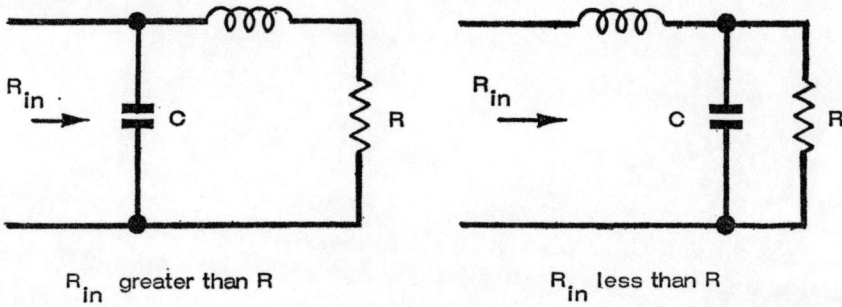

Fig. 81   Impedance matching networks.

diagram shows the circuit which can be used when the desired impedance R_{in} is less than the load resistance R. The following calculations apply:

R_{in} greater than R

$$X_L = \sqrt{R.R_{in} - R^2}$$
$$X_C = \frac{R.R_{in}}{X_L}$$
$$Q \text{ of circuit} = X_L/R \text{ or } R_{in}/X_C$$

R_{in} less than R:

$$X_L = \frac{R.R_{in}}{X_C}$$
$$X_C = R\sqrt{\frac{R_{in}}{R - R_{in}}}$$
$$Q \text{ of circuit} = X_L/R_{in} \text{ or } R/X_C$$

CHAPTER 13

CHECKING RADIO CIRCUITS

A universal meter is the chief tool used for checking or testing transistor circuits. It should be capable of reading voltages up to 25 volts and currents up to 1 amp or 5 amps. For voltage measurement it should have a resistance of at least 20,000 ohms per volt; and for resistance measurements an output terminal voltage of not more than 1·5 volts, as a higher voltage could damage transistors or electrolytic capacitors when the meter probes are applied to the circuit.

A *dc* voltage check across the emitter resistor of each transistor provides an almost complete check on the *dc* conditions and will show up such faults as component failures or open-circuit joints, and also the stage at which the fault is present by the wrong voltage reading at that stage. The fault can then be expected to be confined to components in this stage, which can be checked individually if the fault is not apparent. If individual resistors need checking, it is best to isolate the component(s) for test by unsoldering the connection at one end. If this is not readily possible or desirable, e.g. there may be a number of components to test in the stage, the stage transistor (and the stage) can be isolated by unsoldering the base lead of the transistor. A heat sink should be used in order to avoid heat damage to the transistor and *only* the base lead should be removed from the circuit.

Capacitor faults will not show up on a *dc* check, unless the capacitor is short-circuited. In this case the fault will disappear (as far as *dc* voltage measurements are concerned) as soon as the faulty capacitor is disconnected. By-pass capacitors can be checked for open-circuit faults by temporarily shunting with a 0·5 μF capacitor. If this produces an increase in volume with the set switched on, the original capacitor is either faulty or of wrong value.

The *ac* working of a receiver can only be checked satisfactorily by signal injection. Standard procedure is to feed an audio-frequency into the audio section of the receiver, starting from the output stage and working backwards towards the detector. If all the audio stages work satisfactorily, the *if* stages are similarly tested with the signal generator set to the intermediate frequency modulated with an audio-frequency. Finally the radio-frequency stages can be tested by injecting a radio-frequency signal modulated with an audio-frequency.

In other words, to test the *ac* working of the complete receiver an appropriate signal is injected at each stage in turn, starting with the last stage and working back to the front end. The faulty stage will show up as the first one which does not produce any increase in the audio-frequency output, or yields no final output at all. The necessary signal is usually best injected into the base of the appropriate transistor (i.e. that particular stage transistor), via a suitable resistor in series to ensure current drive. A suitable value of resistor is 10 kilohms. Having found the faulty stage it is then necessary to check the components involved, in turn, in order to determine the source of the fault, e.g. it could be nothing more than a faulty coupling capacitor.

INDEX

ac working, 125, 126
Aerial coils, 15, 16
  coil windings, 24
  coupling, 27, 89
Aerials, VHF, 28
af amplifier, 8–11
Alignment, 99
Alloy diffusion, 49
AM, 9, 21
AM/FM radio, 11
Audio amplifiers, 75 *et seq.*
  amplifier 1-watt, 84
  amplifier 3-watt, 85
  frequencies, 7
Automatic gain control, 97
Autotransformer, 20, 109

Balun, 23
Bandpass filter, 97
Bandwidth, 21
Base, 49, 60
Basic detector circuit, 41
  if amplifier, 96
Battery condition indicator, 122
Bias, 58 *et seq.*, 112
  circuits, 68
  voltage, 59
  voltage, transistors, 58
Bipolar device, 57
Blocking capacitor, 42
'Bottom' end, 113
Breakdown voltage, 36, 37
Bridge circuit, 120
Broadcast frequencies, 14

Capacitive coupling, 103
Capacitor faults, 125
Ceramic filters, 97
Channel current, 113
  voltage, 113
Class A amplifiers, 77 *et seq*
Class B amplifiers, 81 *et seq.*
Coaxial cable, 30
Coding, transistors, 53

Coil design, 15–17
  tapping points, 20, 26
  windings, regenerative, 88
Collector, 49, 59
Control gate, 113
Current amplification factor, 64
  amplification, transistors, 64
  biasing, 69, 72
Crossover distortion, 82
Cross-modulation, 21
Crystal discriminator, 45
  set, 41

Decoupling, 109
Depletion layer, 32
  mode, 113
Detector, 7–11, 39 *et seq.*, 93
Diode characteristics, 34
  rectifier, 120
  temperature effects, 35
Dipole, 9
  aerial, 28
Direct coupling, 104
Directional aerial, 25
Director, 22
Discriminator, 45
Double tuning, 97
Drain, 111
Driver, 81
Dynamic resistance, 18

'Earthy' end, 20
Edge-type meter, 122
Electrons, 32 *et seq.*
Emitter, 49, 59
Enhancement mode, 113
Epitaxial silicon planar transistors, 50
External aerial, 7, 8, 25

Feedback, 68, 87, 109
Feeder impedance, 23
Feeders, 30

Ferrite rod, 24, 89
  rod aerials, 19
  slab, 24
FET, 111 *et seq.*
  amplifier, 44, 86, 115
  transistor, 52
FM, 9, 21
  aerial connections, 30
  detector, 44
Folded dipole, 23
Frequency distortion, 75
Full wave rectifier, 120
Functional groups, transistors, 54

Gate, 111
  voltage, 113
Germanium diodes, 35
  transistors, 49, 54

Half-wave aerial, 28
  rectifier, 120
Hi-Fi FM tuner, 116
Holes, 32 *et seq.*
Horizontal dipole, 29
'Hot' end, 20

if amplifiers, 95
  transformers, 95
IGFET, 113
Impedance, 107–8
  matching networks, 123
Inductance, 15
Inductive coupling, 102, 108
Input characteristics, transistors, 63 *et seq.*
  circuit, transistor, 60, 67
Intermediate frequency, 9, 93
Interstage connections, 103 *et seq.*
Iron dust core, 35

JFET, 113
Junction diodes, 35, 40

L network, 123
lc coupling, 104
Leakage current, 34, 35
Limiter, 45–6
Linearity, 21, 75
Load line, 80, 81
  resistor, 39
Local oscillator, 9–11
  oscillator (mixer), 93
Loft aerials, 22
Long wave, 14
  wave coil, 17
Loop aerial, 25
Loudspeaker power, 75

Medium wave, 14
Mixer, 9, 10, 11
Modulated signals, 21
MOSFET, 113

N-channel, 111
N-P-N transistors, 57 *et seq*

Oscillation, 87
Outline shapes, transistors 55
Output circuits, transistors 60, 67

Parasitic aerials, 22
P-channel, 111
Piezoelectric crystals, 97
Planar process, 49
P-N-P transistors, 57 *et seq.*
Point-contact diodes, 35, 40
Polarities of transistors, 57
Polarized signal, 28
Positive bias, 33
Potential barrier, 33
Power gain, 78
  supply, 119, 120
  transistors 55
Preamplifier, 86, 90, 94

Pre-selector, 31
Pro-Electron coding 54
Pulse counting detector, 45
Push-pull circuits, 81 *et seq.*

Q factor, 17, 18, 20, 87
Quality factor, 13, 15

Radio frequency, 9
Ratings, transistors, 66
Ratio detector, 44, 45
RC coupling, 104
Reactance, 18, 109
Rectifier, 32
Reflector, 22
Regenerative circuit, 8
  receivers, 87 *et seq.*
Reservoir capacitor, 39
Resistance of transistors, 67
Resonance, 13
Resonant aerial length, 31
  frequency, 13
Reverse bias, 33
rf amplifiers, 86
Ribbon aerial, 23

Secondary Current, 107
  voltage, 107
Self-oscillating mixer, 94
Selectivity, 19, 20
Sensitivity, 20, 97
Short wave, 14
  wave coil, 17
Signal gate, 113
  generator, 99
Silicon diodes, 35
  transistors, 49, 54
Source, 111
Specification figures,
  diodes, 38
  transistors, 66

Superhet, 10
  alignment, 99 *et seq.*
  receivers, 93 *et seq.*
Symbols, transistors, 60

Telescopic aerial, 29
Temperature effects, diode, 35
Thermistor, 82
Time constant, 40, 104
Tone control, 118, 119
'Top' end, 113
Transfer characteristics, transistors, 64
Transformer coupling, 102
Transformers, 106–7
  if, 95 *et seq.*
Transistor outline shapes, 55
  parameters, 62
  power, 78
  ratings, 66
Transistors, 48 *et seq.*
Trimmers, 19, 102
Tuned circuit, 7, 13
Tuning capacitor, 15, 16
  meter, 121
Turns ratio, 106
Types of diodes, 35

Universal meter, 99, 125

Variactor, 36
Varicap, 36
VHF, 9, 14
  aerial, 28
Voltage biasing, 70
Voltage-stabilizing, 36
Volume control, 43, 86

Wire aerial, 25–8

Zener diode, 36, 54